The Institute of Biology's Studies in Biology no. 142

Plasmids

Martin J. Day
B.Sc., Ph.D.
Lecturer in Applied Biology, University of Wales Institute of Science and Technology, Cardiff

Edward Arnold

First published 1982
by Edward Arnold (Publishers) Limited
41 Bedford Square, London WC1 3DQ

British Library Cataloguing in Publication Data

Day, M. J.
Plasmids.—(The Institute of Biology's studies in biology, ISSN 0537-9024; no. 142)
1. Plasmids
I. Title II. Series
589.9'08734 QR76.6

ISBN 0-7131-2846-1

Photoset and printed by Photobooks (Bristol) Limited

General Preface to the Series

Because it is no longer possible for one textbook to cover the whole field of biology while remaining sufficiently up to date, the Institute of Biology proposed this series so that teachers and students can learn about significant developments. The enthusiastic acceptance of 'Studies in Biology' shows that the books are providing authoritative views of biological topics.

The features of the series include the attention given to methods, the selected list of books for further reading and, wherever possible, suggestions for practical work.

Readers' comments will be welcomed by the Education Officer of the Institute.

1982

Institute of Biology
41 Queen's Gate
London SW7 5HU

Preface

Studies of plasmids have revealed them to be widespread in bacteria. I have tried to convey to the reader some of the experimental approaches which have been used in these studies. We now have an appreciation of features of their natural history even though we are still far from understanding their ecological roles. The aim of this book is to give to the reader an insight into the biology of plasmids. I hope the straightforward account given will stimulate interest into this intriguing area of research and allow the reader to acquire a 'feel' for a rapidly developing area of biology.

Cardiff, 1982

M.J.D.

Contents

1 What are Plasmids?

1.1 Opening remarks

A major milestone in the study of bacteria was the recognition that they were far from being just simple, asexually reproducing micro-organisms. Bacteria were found to possess a parasexual cycle, similar in some respects to the obligate sexual cycles observed in higher organisms, and subsequently shown to undergo recombination and gene rearrangements implying that gene flow could occur in bacterial populations. The history of plasmid biology started in 1952 when Lederberg and his colleagues discovered that the plasmid sex factor F was capable of transferring genes from the host bacterium to a recipient bacterium. Later, in 1960, Mitsuhashi and co-workers showed this academic study to be both conceptually and practically analogous to infectious transferable drug resistance agents. These are termed R factors.

1.2 A definition

Plasmids are extra pieces of DNA, mini-chromosomes, which can replicate independently of, and coexist with, the host chromosome. Since *autonomy* is not an inherent feature of every piece of DNA, those duplexes, i.e. plasmids and chromosomes, capable of maintaining themselves in the cell, are termed *replicons*.

Under most conditions of growth, biosynthetic genes, e.g. those coding for amino acid biosynthesis, are essential for cell viability. These are found almost without exception on the chromosome. In contrast, naturally occurring plasmids often carry genetic information which is essential for cell growth under 'abnormal cultural conditions'. Plasmids generally maintain genes which code for structures or chemical transformations that may, under certain conditions, be of critical survival value to the host cell. For example the presence of an *R factor*, that is a plasmid which carries antibiotic resistance genes, will enable the host cell to survive successfully a challenge from those antibiotics. Another feature of many, but not all, plasmids is that they are self transmissible. There are two transfer mechanisms, *conjugation* and *transduction*. Some plasmids possess genes which enable them to synthesize a *pilus* which is a *proteinaceous* tube-like structure that is extruded from the cell and, on recognizing a recipient cell, binds to it. When cell-to-cell contact is established a copy of the plasmid is transferred, through this pilus, into a recipient cell (see section 1.5). Plasmids with this conjugal capability are termed *sex factors*. The second mechanism of plasmid transfer has been called transduction (see section 1.5). Virus genomes possess many structural features in common with other replicons. To distinguish between viruses and plasmids and treat either in isolation would be artificial and to suggest that they have some profound differences. In the past they have been treated separately which had led to confusion about their relationship and identity. Virus genomes contain genetic information for the viral protein 'coat' which has some similarity to pili. Each virus genome is packaged in its 'coat' and subsequently released from the cell. The virus coat protects the DNA carried within. Once it recognizes a suitable recipient, binding or 'docking' occurs at a particular site on the outside of the bacterial cell and the viral DNA is injected into the cell. The mechanism does not require cell-to-cell contact but nevertheless shows similarities to conjugational gene transfer. In most cases the genes transferred in the virus coat are those of the virus; however, there are many examples showing that other plasmids and even segments of the host's chromosome can be transferred. The process which results in the transfer of non-viral DNA is termed *transduction*.

The replication and maintenance of extrachromosomal DNA creates an energy demand on the host cell. Since the advantages of a plasmid to the host cell are of intermittent value, cells that do not have this replicon would be expected to be at a distinct growth advantage in normal growth conditions. From this synopsis it might be predicted that plasmids would have a peripheral role to play in the life of bacteria. This prognosis will be examined in more detail in Chapter 8.

1.3 Structure and physical characteristics

The bacterial chromosome and plasmids have been shown to exist as double stranded closed covalent circles of DNA (*CCC DNA*). The two replicons differ in size; in general plasmids are 0.01%–2% (about 1–200 megadaltons) of the chromosome. Plasmid size is variously measured in kilobases (kb), megadaltons (Md) and molecular weight (MW). There are about 10 kb to 6.5 Md and Md is synonymous with MW. The lower size limit is set by one essential function, the ability to replicate independently of the chromosome.

1.3.1 Preparation from cells

Plasmids can be prepared by a variety of techniques all of which begin with gentle lysis of cells. The plasmid normally exists in a highly twisted or supercoiled configuration. It appears that during replication one of the strands of the duplex is cut and it is then no longer *CCC DNA*. As a result of this a plasmid preparation naturally consists of two types of molecule, one twisted (*CCC*) and the other *relaxed* (*OC*), (Fig. 1–1). During preparation, both

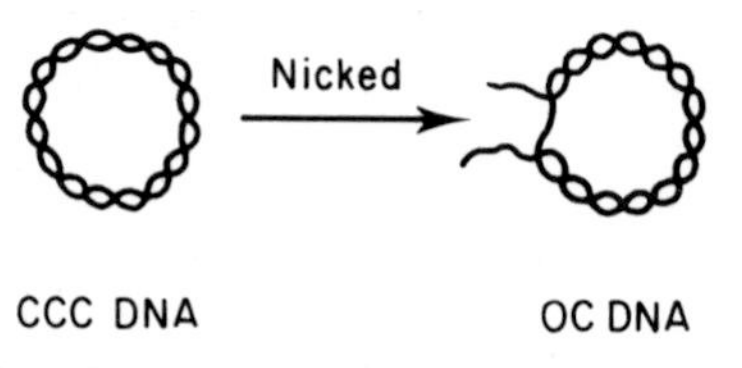

Fig. 1–1 The molecular forms of plasmid DNA.

chromosome and plasmid are subjected to shearing forces and these increase the amount of relaxed plasmid DNA found as well as reducing the chromosomal DNA to linear strands.

1.3.2 Purification

Plasmid *CCC DNA* can be separated from linear chromosomal DNA by centrifugation. Ethidium bromide is a dye which binds to, or *intercalates* between, the DNA bases. As a result, both linear and chromosomal DNA and 'relaxed' plasmid DNA open further and bind more dye. This increases their buoyant densities. The highly twisted state of the *CCC* plasmid DNA imposes a physical constraint on the total amount of dye bound and therefore the plasmid remains compact. As a direct result, under *centrifugation* in a caesium chloride (CsCl) density gradient, the plasmid fraction separates from DNA in other states (Fig. 1–2). Recently other techniques have been reported for the isolation and purification of plasmid DNA which involve the use of *chromatography* and *gel electrophoresis*.

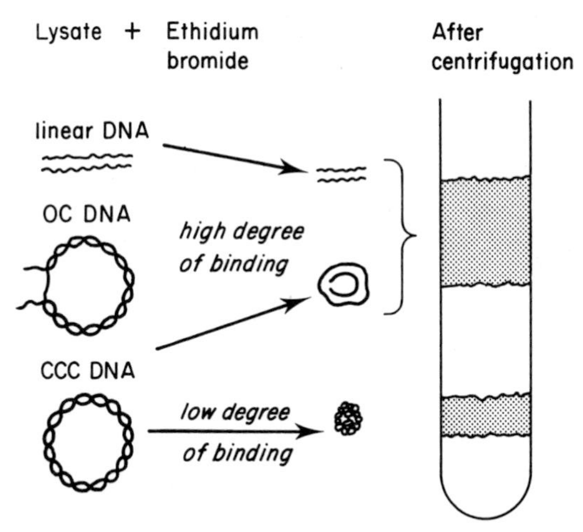

Fig. 1–2 Plasmid preparation by centrifugation in an ethidium bromide/caesium chloride density gradient.

1.3.3 Physical analysis

Marmur and his colleagues were the first to exploit the affinity that *homologous* or *complementary sequences* of DNA have for one another. A heated DNA duplex denatures, or *melts*, into single strands. When cooled in suitable conditions of salt and pH it spontaneously reforms or *anneals* into the original duplex structure. A procedure based on this technique is called *heteroduplex mapping* and can be used to detect differences in homology of as little as 100–200 base pairs between two plasmids. Such a comparison between parent and mutant or related plasmid can reveal precise regions of homology and difference

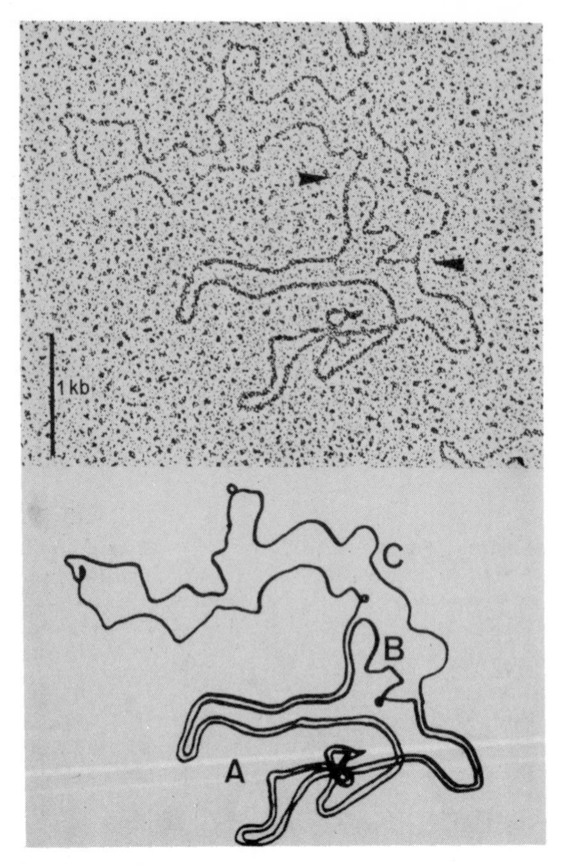

Fig. 1–3 A heteroduplex molecule between two plasmids of different size (12.8 kb and 21.5 kb). The plasmids share an 11 kb common sequence (A) which form a double strand. The remaining 1.8 kb segment (B) of the smaller plasmid is non-homologous to the 10.5 kb segment (C) of the larger. The arrows indicate the branch points. (Courtesy of Dr J. Meyer, Biozentrum, University of Basel, Basel.)

(Fig. 1–3). The examination of electron micrographs of a heteroduplex is, therefore, a powerful tool for mapping plasmids.

The enzymes known as *restriction enzymes* cleave DNA at specific sites. Each enzyme has a unique base sequence which it recognizes as its substrate. One or more such enzymes can be used to digest a plasmid preparation. When the digest is subjected to gel electrophoresis a unique set of bands, known as a 'fingerprint' is produced (Fig. 1–4). The number of fragments in a plasmid digest depends on the position and number of substrate sites for the restriction enzyme chosen. The fingerprint will therefore vary depending on the choice of plasmid and restriction enzyme. Furthermore, this digest pattern can be changed by deletions, additions or mutation to the plasmid DNA, altering or changing the relative positions of restriction sites.

Both these techniques are used to identify relationships between plasmids.

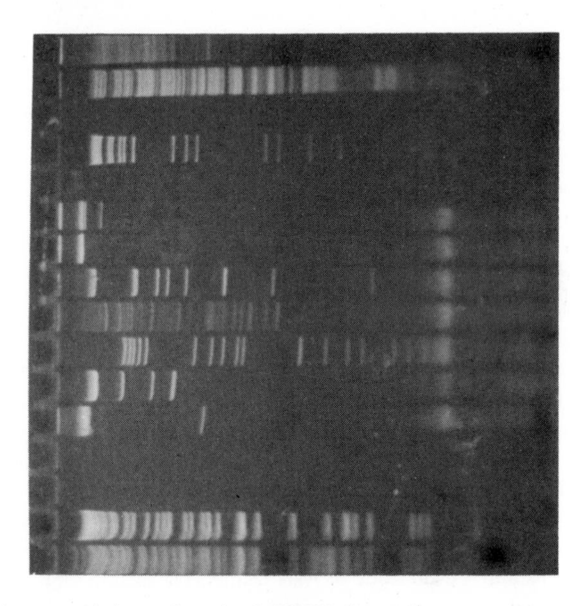

Fig. 1–4 Agarose gel electrophoresis of pBR322 digested by seven restriction enzymes. The number of bands in a channel is an indication of the number of restriction sites present in the plasmid DNA for that enzyme. The electrophoresis was carried out from left to right, the smaller fragments migrate the furthest. (Courtesy of Mr K. Ineichen, Biozentrum, University of Basel, Basel.)

1.4 Genetic identification

1.4.1 Phenotype

A plasmid is termed *cryptic* if it confers no recognizable phenotype on the host cell. Plasmids are recognized genetically by the abilities or phenotypes they confer on the host cell (Table 1). These phenotypes have been recognized as a diverse collection of characters of which many confer a growth advantage to the host cell under 'abnormal growth conditions'. In nature, bacterial populations are subjected to a fluctuating environment and this might, in part, explain the observed diversity in plasmid types and characteristics.

Included in this range are R factors, catabolic plasmids and viruses. All have their particular niches, but share many common features and as such should be treated as a variation on a theme under the term plasmids. How does one recognize that a particular character is plasmid borne? Most plasmids endow their host cell with a recognizable phenotype although for others, for example viruses, the phenotype is less obvious. Virions can be recognized in the electron

Table 1 Some properties of naturally occurring plasmids.

Property	*Plasmid*	*Host*
Fertility (the ability to transfer genes by conjugation)	F	*Escherichia coli*
Colicinogenicity	ColE1	*E. coli*
Antibiotic production	SCP1	*Streptomyces coelicolor*
Heavy metal resistance (mercury)	FP2	*Pseudomonas aeruginosa*
Ultraviolet resistance	ColIb	*E. coli*
Enterotoxinogenicity	Ent	*E. coli*
Virulence (haemolysin)	Col V	*E. coli*
Catabolism (e.g. camphor)	CAM	*P. putida*
Tumourgenicity	T-1	*Agrobacterium tumifaciens*
Restriction/modification	R1	*E. coli*
Drug resistance (penicillinase)	RP1	*P. aeruginosa*
Protein coat (Virus)	Lambda	*E. coli*
Confers host cell sensitivity to virus MS2	F	*E. coli*

microscope but are rarely assessed this way. The recognition of a virus is usually made through their ability to induce a *plaque* in a lawn of sensitive cells. A plaque is a circular clearing in a turbid bacterial lawn, often about 2 mm in diameter, and varies in size depending on the virus, host and cultural conditions. The plaque is formed from the infection of a bacterium by a virus which proceeds to multiply and lyse the host cell. The freshly released virus particles then infect cells and a clear zone of lysis results (see section 1.5).

The host range for plasmids is also variable. For instance some plasmids, F and λ, have a very limited host range, restricted to a few *E. coli* strains. Whereas RP1 has a wide host range, it will transfer to all Gram-negative bacteria tried and bacteriophage P1 is capable of infecting several species of enterobacteria, including *Escherichia coli, Klebsiella aerogenes* and *Salmonella typhimurium*.

Further genetic proof of identity can be obtained by showing that the plasmid borne characters are associated with the events of curing, incompatibility and surface exclusion.

1.4.2 Curing

A general observation of plasmid borne characters is that they are unstable, i.e. spontaneously lost or *cured* at an observable rate. The rate of curing can be as low as the mutation rate for a gene in the host, but in many cases it occurs at a much higher frequency. Once such a character is lost it cannot be retrieved by back-mutation. For example, when an R factor bearing strain (R^+) is cultured in the laboratory in a non-selective nutrient medium, the genes for drug resistance are frequently and spontaneously lost. This loss can be due to elimination of either the entire plasmid or just the resistance genes themselves from the plasmid. This loss is due to the failure of the plasmid to replicate 'in step' with cell division. Any cell which accidentally fails to inherit a copy of the plasmid will presumably have a reduced energy expenditure per generation and under these 'non-selective' conditions, will generally outgrow the R^+ strains. Spontaneous plasmid loss is highly variable and depends on both environmental factors and plasmid-host interactions; consequently it is an unpredictable event. In research, techniques have been developed aimed at increasing this level of instability in order to prove that characters are plasmid borne. This also generates plasmid free strains which can then be used as recipients in back crosses. The techniques used to promote curing include exposing the host strain to a variety of cultural conditions, e.g. ethidium bromide, acridine orange, mitomycin C, sodium dodecyl sulphate, urea, U.V. irradiation and temperature shock. The efficiency of any one of these methods depends again on both the plasmid and the host cell.

There are two different modes of curing exhibited by these agents. Firstly, the dyes, U.V. irradiation and temperature shock: these all interfere with plasmid replication and hence cause plasmid loss through segregation at cell division. Secondly, the surface active agents do not affect plasmid replication but cause pili to dissociate. This prevents cells which have spontaneously lost a plasmid from being reinfected. Such curing agents are therefore only effective against plasmids possessing sex factor activity. Curing is not always the 'all or nothing' process outlined above. A plasmid often possesses several characteristics, e.g. resistance to streptomycin, carbenicillin and tetracycline, and these can frequently be shown to be lost independently as well as coincidentally. The term *dissociation* describes this process of character segregation (see Chapters 4 and 5).

It should be noted that several of these curing agents are also mutagenic. 'Cured' strains must, therefore, be screened to ensure that they are not mutants in those plasmids phenotypes being examined.

1.4.3 Incompatibility

Cells can acquire plasmids by the processes of conjugation, transduction and transformation. A resident plasmid can be displaced and lost from the host cell on acquisition of a new plasmid: such an occurrence is known as *incompatibility* and demonstrates a relationship between the plasmids. This loss is thought to be due to competition between the two plasmids for the same replication site. The plasmid which cannot occupy this site cannot replicate and segregates at cell division. It is, therefore, lost from one cell line.

1.4.4 Surface exclusion

The presence of a plasmid, whether it be a virus or sex factor is often sufficient to confer immunity against *superinfection* by further virus or plasmids using the same adsorption sites. This *surface exclusion* property is distinct from incompatibility. It appears to be a cell surface phenomenon produced by the resident plasmid. It is possible that there is some structural change in, or covering up of, the sites for phage or pilus attachment. If pili or phage cannot recognize attachment sites then the process of infection cannot proceed. This interference with adsorption reduces, and can eliminate, their entry.

In summary, therefore, there are three genetic characteristics which imply a plasmid location for a phenotype. Firstly a high rate of curing of the suspected plasmid characteristics. Secondly an ability to promote or inhibit the transfer of these characters between cells by conjugation or transduction. Finally the loss of these characters on the entry of a second incompatible plasmid. It is important to tie both the physical and genetic evidence together for conclusive proof of plasmid involvement.

1.5 Plasmid life cycles

Three examples have been chosen to illustrate some of the key features in the life cycle of plasmids. They have been extensively analysed and reported, and form both conceptual and practical bases for much of the work carried out in plasmid research.

1.5.1 Lambda

λ is a typical temperate virus. It is potentially lethal, but often co-exists in a quiescent state *integrated* into the host cell chromosome and termed a *prophage*. The host cell is now termed a *lysogen*. The virus can spend considerable periods integrated before becoming activated. When the virus genome is activated it *excises* from the chromosome and proceeds to replicate. Many mature viruses are formed in each host cell, which then lyses. These free virus can then attach and reinfect a susceptible host. It is possible for λ to transfer chromosomal genes, by the process of *transduction*. This is achieved by the virus genome

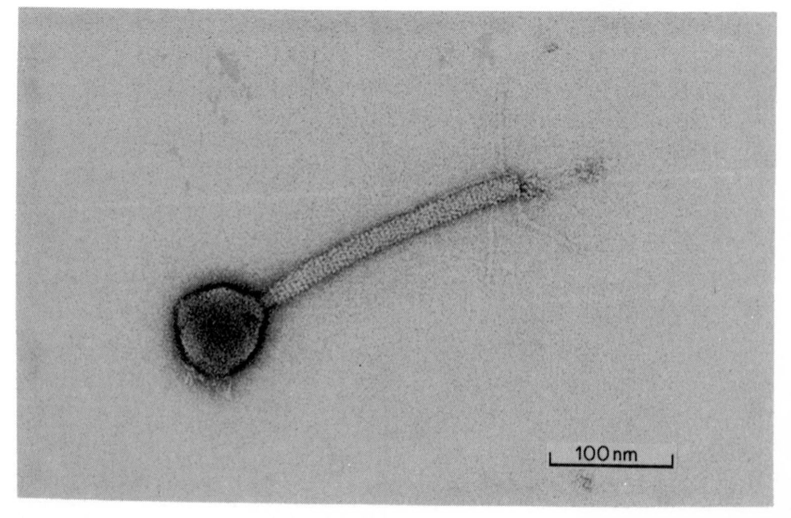

Fig. 1–5 An electron micrograph of bacteriophage P1. (Courtesy of Dr B. Heggeler, Biozentrum, University of Basel, Basel.)

excising from the host chromosome in a faulty manner. When this occurs the virus genome can gain some of the adjacent bacterial genes leaving some of its own genes on the chromosome. This *hybrid* genome is packaged in the viral coat and when released it is able to infect a new host cell. On infection the transduced bacterial genes can then replace by *recombination* the complementary chromosomally located genes. The recombination is carried out by an enzyme coded for by the *rec*A gene. This technique has been used extensively for gene mapping in bacteria.

1.5.2 Sex factor F

The F factor is by all criteria a benign plasmid, but it is able to modify the host cell, which is called an F^+ or male, into a vehicle for conjugational gene transfer. This plasmid carries genes resulting in the production of a pilus which can be distinguished from other pili by the ability of phage MS2 to adsorb to it and infect the cell. On 'docking' one strand of the plasmid duplex is transferred through the pilus (Fig. 1–6). Both strands are replicated in their respective host

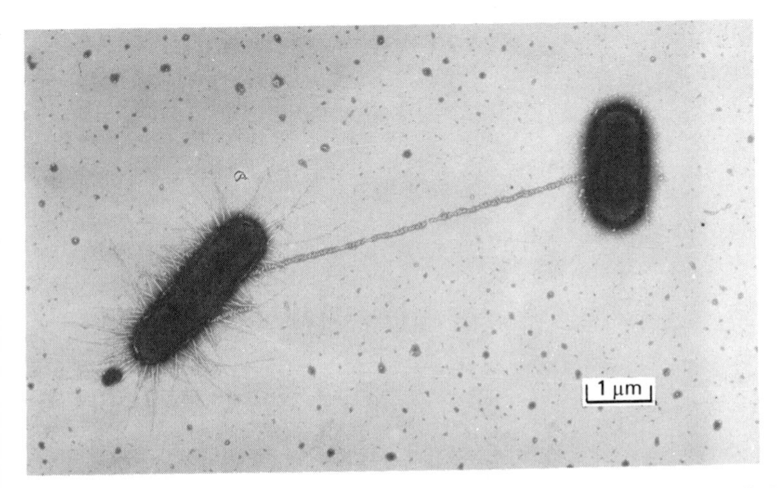

Fig. 1–6 An electron micrograph of conjugating *E. coli* cells. The F pilus is labelled along its length by virus particles which specifically infect the donor cells, through their pili. (Courtesy Prof. C. C. Brinton Jr., University of Pittsburg, U.S.A.)

cells and the net result is two male cells. The process is termed *conjugation* and is achieved by plasmids which have *sex factor activity*. The plasmid F, like λ, can integrate into the host chromosome. In this state it is able to transfer chromosomal genes to female cells at high frequency. Strains with F integrated are known as *Hfr* (*high frequency of recombination*) *strains*. Jacob and his colleagues (1960) suggested that plasmids with the ability to integrate be termed *episomes*. Such elements replicate either autonomously or under chromosomal control when integrated.

1.5.3 Bacteriophage P1

P1 is a temperate bacteriophage (Fig. 1–5) similar in many respects to λ. There is one major difference, however, P1 rarely integrates into the host chromosome. In a lysogenized cell P1 prophage exists independently of the chromosome as a replicon. This difference has led to P1 being termed both a plasmid and a bacteriophage. However, it is possible to obtain mutants of λ that can exist in this autonomous state and mutants of P1 that cannot lysogenize. This demonstrates an obvious interrelationship between the two elements and it therefore seems artificial to separate them in this manner.

In this chapter it has been shown that plasmids cannot be defined in a rigid manner because they are, by their very nature, a labile and diverse collection of genetic elements. In the following chapters we explore in more detail these aspects of plasmid biology.

2 Plasmid Gene Transfer

2.1 Processes of plasmid gene transfer

There are three possible mechanisms by which plasmid transfer can be achieved and these have been described as conjugation, transduction and transformation. To understand these processes and the factors affecting them it is important to be familiar with the principles and methodology which have been used in their study.

2.1.1 Conjugation

Conjugation is achieved experimentally by the use of two genetically identifiable strains. For example, the donor strain is sensitive to tetracycline (Tc) but carries a drug resistance plasmid which could specify resistance to several drugs, e.g. sulphonamide (Su), streptomycin (SM) and pencillin (Pen). Conversely the recipient is resistant to Tc, due to a chromosomal mutation, but is sensitive to the other three drugs. Cultures of the two strains are mixed and incubated without shaking. A sample is taken immediately, agitated violently, to prevent pairs of cells mating and then plated on to selective medium containing all four of the drugs (neither donor nor recipient is able to survive independently). Consequently colonies would not be expected to arise at this stage because mating has been prevented. Samples are then withdrawn at intervals and treated in a similar manner to the initial sample. Selection by the medium is thus made for recipients which have acquired the R factor by conjugation and subsequently drug resistant colonies (resistant to all four antibiotics) grow on the medium. Figure 2.1 shows that the R factor first 'appears' in recipient cells, termed *exconjugants*, after 10 minutes, the numbers then increasing rapidly to a threshold some 10 minutes later. This plateau is equivalent to transfer from some 90% of the donor population. The efficiency of conjugal transfer of the R factor to a recipient population is obviously dependent on the donor/recipient ratio, the total number of cells involved in the mating and the ability of cells to act as donors and recipients.

This experiment also shows that R factor borne genes are transferred in matings as a 'unit' and that cell-cell contact is required for transfer. Both pH and temperature of the medium have an effect on conjugation (Fig. 2–2a and b). In the example chosen the optimal pH is 7.0 and the optional temperature for R factor transfer is about 30°C. Changes in these and other factors, e.g. oxygen tension, viscosity and salinity can greatly affect transfer. Other plasmids may have different optima.

A second feature of this type of gene transfer is indicated by there being no requirement for recombination, since the R factor is autonomous. In theory, one

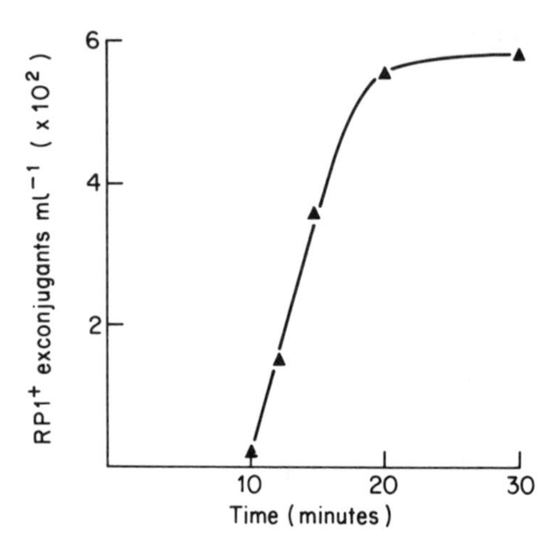

Fig. 2–1 Conjugal transfer of RP1.

would expect the entire population of cells ultimately to acquire the R factor, but this is rarely achieved because a certain number of the putative recipient cells are incapable of conjugation. This could be due to their being in an unsuitable physiological state or because they possess mutations which affect either their ability to act as recipients or to maintain the plasmid. Since the efficiency of the conjugal process depends, in part, on the establishment of cell-cell contact it is not surprising to find that plasmid transfer is impossible in cases where the pili cannot recognise or 'dock' with recipient cells (see section 2.4). This 'docking' specificity partly explains the basis of host range in plasmids.

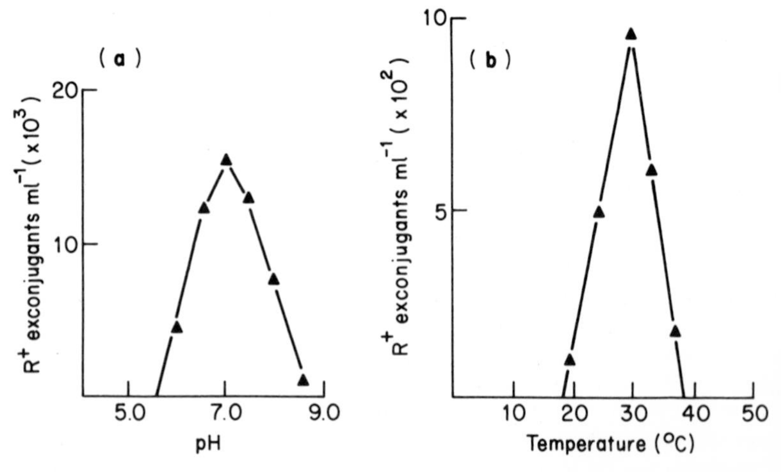

Fig. 2–2 **(a)** Optimal pH of the medium (at 30°C) for conjugal transfer of an R factor. **(b)** Optimal temperature.

2.1.2 Transduction

The 'docking' concept applies equally well to this second mode of transmission (see section 1.5.1). When virus particles are released from a lysed cell they contain genetic information. It is the protein coat of the virus alone that is instrumental in the stages of host recognition, attachment and injection of the DNA carried. The efficiency of this process relies on the ability of the bacteriophage to recognize receptor sites on the bacterium cell surface.

It is obvious that an inability to meet with the next host cell renders both conjugation and transduction impossible. Both these have been extensively investigated through laboratory experimentation. Results have shown that equally high rates of gene transfer can be achieved. Some plasmids are very efficient at gene transfer whilst others are not. Furthermore, the ability to transfer genes by conjugation can often be repressed by adverse physiological conditions which result from a poor growth environment. A similar effect can be seen for some viruses which, although capable of generating transducing particles, are prevented from initiating their lytic cycle until suitable growth conditions return. This also ensures that growth conditions are optimal for potential host cells.

2.1.3 Transformation

This process, whereby 'naked DNA' is taken up by the host cell, has been regarded as having little ecological significance. In the near future however it may be shown to contribute to gene transfer in particular habitats. For instance there is evidence for both intraspecies and interspecies transformation in both *Pneumococcus* spp. and *Streptococcus* spp. Evidence is accumulating to show that many bacteria are capable of taking up external DNA. These observations suggest that bacteria, some of which carry plasmids, release this DNA by lysis or excretion and hence make it available for 'uptake' by suitable organisms. It may be that a certain proportion of the population is sacrificed to make this genetic information available to the remaining population. The transformation of plasmids might be expected to be successful because, intracellularly, many plasmids exist as stable autonomous units. Upon acquisition they do not need to be integrated into the host chromosome to be stabilized.

2.2 Restriction and modification

As the chromosome of *E. coli* replicates, a modification enzyme attaches a methyl group to one specific base sequence on the DNA wherever it occurs. Two bases in this sequence, one on each strand, are methylated and the chromosome is then said to have undergone *modification*. Genes specifying enzymes with this ability are found on both the bacterial chromosome and on some plasmids. The cell also possesses a restriction system which can recognize unmodified DNA. This system is mediated by enzymes known as *restriction endonucleases*, which specifically attack DNA. They can 'nick' or 'cleave' both strands at the phosphodiester bond in the DNA helix. Each system is unique in that restriction and modification is achieved through one specific base sequence. EcoR1 endonuclease will attack and cleave a sequence (Fig. 2–3) at the two points indicated (↑) unless one or both of the strands are modified. A common feature of restriction sites is their symmetry. Sites with twofold symmetry are termed *palindromic* sequences. What then is the significance of this highly ordered

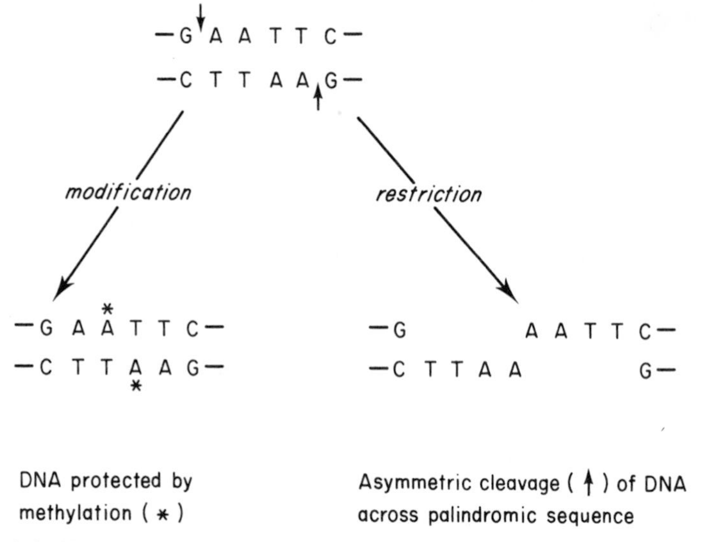

Fig. 2–3 The cleavage site of EcoR1 restriction endonuclease and its protection by methylation.

process? Modification labels the host DNA in a manner that enables it to be easily and efficiently distinguished from non-host DNA arriving by gene transfer. The restriction enzyme will survey incoming DNA inspecting it for, and attacking, unmodified sites. There are many restriction systems known for *E. coli* and other bacterial species. Possession of a system for restriction and modification confers several advantages on an organism; such a system facilitates speciation and the efficient colonization of ecological niches. Genetic isolation also permits specialization and consequently imported genes from similar genetic backgrounds are more likely to work and be stable in such backgrounds. The occurrence of restriction and modification systems suggests that the flow of genes may be confined between individuals and populations of cells that are genetically related.

2.3 Phenotype and host range

This aspect was discussed in some detail in section 1.4. The transfer of an R factor from one bacterial species to another may dramatically change the level of drug resistance expressed by the plasmid. For example Table 2 shows that, for

Table 2 The drug resistance levels exhibited by bacterial species infected with the same R factor (carrying sulphonamide, streptomycin and ampicillin resistances).

R^+ *strain*	*Minimal inhibitory concentrations* (μg ml^{-1})		
	Sulphonamide	*Streptomycin*	*Ampicillin*
E. coli	1000	20	250
Shigella sonnei	1000	200	250
Salmonella typhimurium	1000	125	100
Proteus mirabilis	1000	100	25
Klebsiella pneumoniae	1000	125	250
Resistance prior to the acquisition R factor	1–50	0.3–10	3–7.5

the same R factor, streptomycin resistance levels are lower in *E. coli* than in *Salmonella* spp., *Shigella* spp. and *Klebsiella* spp.

These differences in levels of resistance are significant, and suggest that the host cell may lack essential structures or suitable cofactors for the optimal expression of the resistance genes. Even so the resistance levels conferred on the host by this R factor are frequently high enough to preclude the clinical use of these antibiotics.

Table 3 shows the transfer frequency of R74 from a clinically isolated *E. coli* to a variety of recipients. The table illustrates the differing transfer efficiencies that can occur with a broad host range, or *promiscuous*, plasmid.

Table 3 The transfer frequency of R74 mediated carbenicillin resistance from a clinically isolated *E. coli* to a variety of recipient strains.

Recipient	*Frequency of carbenicillin transfer*
P. aeruginosa	2.10^{-6}
E. coli	6.10^{-3}
S. flexineri	1.10^{-1}
S. typhimurium	3.10^{-7}

Transfer frequencies were measured by mixing growing donor and stationary phase recipients (50:1). After 15 minutes the mating was stopped by plating the cells on selective medium.

2.4 Factors affecting plasmid distribution

Plasmids are widely spread in natural populations of bacteria, and individual bacteria can act as host to a large number of these elements. This is illustrated by one report of a pseudomonad harbouring 13 different viruses. There is now little reason to suspect that this is one end of a spectrum that exists in bacterial populations. The spread and maintenance of plasmids is probably modified by several interactive processes. One of these acts at the cell surface and is termed surface exclusion (section 1.4.4). A second arises through the interaction between two plasmids which cannot co-exist and therefore segregate at the division of the cell. This is termed incompatibility (see section 1.4.3). A third refers to the inhibition of fertility by compatible plasmids.

2.4.1 Fertility inhibition

If one of two compatible plasmids has sex factor activity it is possible for a second plasmid, which could be another sex factor, a non conjugal element, or a virus, to inhibit the fertility or conjugal ability of the first. This can be a reciprocal ability and suggests that the two elements have a common control mechanism affecting fertility. It is often achieved through the inhibition of pili synthesis. In *E. coli*, plasmids that inhibit the fertility of F were previously termed fi^+ and those that have no effect fi^-. The incompatibility (inc) classification system has largely superseded this scheme (fi^+/fi^-).

In summary the full significance of conjugational gene transfer in microbial populations is as yet not fully appreciated. Some recognition of its roles in medical microbiology and microbial ecology has been made. Although the frequency of viruses is high in natural systems, the proportion of those capable of transduction is not known. Even less is known of the amount, or significance, of gene transfer in natural populations of bacteria, by transformation or transduction.

3 Bacteriocins

Many bacteria produce diffusible protein antibiotics, known as *bacteriocins*. These were recognized as having a lethal effect on other bacteria of the same or closely-related species. They have been thought of as distinct from antibiotics because of this limited range in activity. However recent work has demonstrated that some of these agents are also effective against completely unrelated species.

The ability of *E. coli* strains to produce *colicins* (bacteriocins produced by *E. coli*) active against other enteric bacteria was first documented by Gratia and Fredericq in 1946. Genes for the synthesis of these proteinaceous agents are frequently plasmid borne. There are two principal types of colicin or Col plasmid, those with conjugal ability (MW $60-100 \times 10^6$) and smaller ones lacking this ability (MW about 5×10^6).

Bacteriocins have been shown to occur in many bacterial genera and are probably universally distributed. Since most of the work to date has been carried out in *E. coli*, colicins will form the major part of this discussion.

3.1 Colicin recognition

Suspected colicinogenic strains are spotted on the surface of a nutrient agar plate and incubated for 24 hours at 37°C. The plate is then inverted, a chloroform soaked paper disc placed in the lid and the vapour allowed to kill the bacterial colonies. The surface of the plate is then overlaid with molten agar seeded with a sensitive strain of *E. coli*. After overnight incubation zones of clearing can be observed. These signify that the test strains may have produced a bacteriocin. Such zones could also be produced by a virus but these can be distinguished by placing the point of a needle in the cleared zone and transferring to a fresh lawn of the sensitive strain. If a virus is present another zone will develop, whereas colicins will not form a new zone.

However, many bacteria produce both bacteriophage and bacteriocins. Phage activity, but not the colicin activity, can be inactivated by irradiating with ultraviolet light and this enables which ones are present to be ascertained. There are still other factors which may interfere with this analysis. These include the production of secondary growth compounds, lytic enzymes, hydrogen sulphide, toxins and acids, all of which could cause zones of inhibition similar to those of lysis.

The recognition of colicins by bioassay is further complicated by Col^+ strains that produce only small amounts of colicin. In culture it seems that only 1–10% of cells actually produce the colicin and that this is age dependent. As the culture approaches the end of its growth phase, the amount of colicin present is seen to increase. This production coincides with the lysis of a portion of the population.

It has been suggested that this is altruism; a sacrifice of a minority of the population releasing the colicin which may inhibit sensitive competing species. This hypothesis has been challenged recently by a suggestion that colicins are not produced by colicinogen strains in anaerobic environments e.g. in the intestine. The amount of colicin produced can be increased by temperature shock, irradiation and a variety of chemicals.

3.2 Identification

Colicins have been classified on the basis of their activity against a set of colicin resistant mutants. Such resistant mutants can be easily isolated since they form discrete colonies inside the zones of clearing (see Fig. 3–1). Most mutants are insensitive because they have lost their receptor for binding the colicin. Some

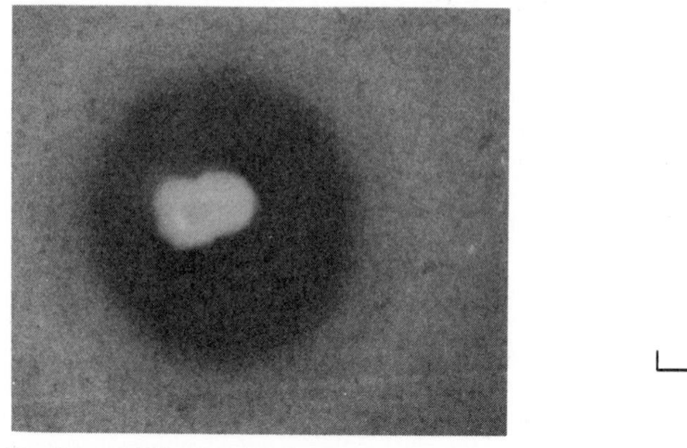

Fig. 3–1 Colicin E3 production by a single colony. The colony is killed and overlaid with a sensitive *E. coli* strain. Colicin-insensitive mutants appear within the inhibition zone. (Courtesy of Dr A. Pugsley, Biozentrum, University of Basel, Basel.)

colicins are indistinguishable by this assay, e.g. colicins E1 and E3 because they bind to the same receptor site. However, a bacterium harbouring ColE1 is resistant to colicin E1 but is still sensitive to colicin E3 produced by ColE3 and vice versa. This second assay system, which is based on immunity, further aids the identification of colicins.

3.3 Colicin structure

Colicins are proteins with a molecular weight of 27–90 000. Some show similarity and this suggests a common ancestry. Colicinogenic plasmids fall neatly into two groups, as shown in Table 4, and can be classified in the same manner as other plasmids. For example, a group 2 Col plasmid, ColV–K94, has been shown, by heteroduplex analysis, to have a 40% identity with the fertility factor F. Similarly, when the plasmid ColIb is present in the same cell as F it reduces the conjugative ability of the F factor.

Table 4 The characteristics of Col plasmids.

Group 1	Group 2
ColE1	ColIb
MW 5×10^6	62×10^6
about 10 proteins	about 100 proteins
about 10 copies $cell^{-1}$	1–2 copies $cell^{-1}$
non-conjugative	conjugative
(transferable by other plasmids)	

A conceptual problem arises as this type of investigation proceeds. How can a bacteriocin be defined and distinguished from an antibiotic? Bacteria produce a wide range of proteinaceous agents, many of which we recognize as antibiotics. More recent research has demonstrated another class of these agents which are also plasmid borne, and these have been termed microcins. These have been observed in both *E. coli* and *P. aeruginosa*. Microcins are small, having a molecular weight of about 1000, and are effective against a variety of bacterial species. In some cases methionine has been found to be antagonistic, suggesting that the site of action is concerned in some way with the metabolism of this amino acid. These are very different to colicins, which have a high molecular weight (10^5) and the peptide antibiotics, which have low molecular weight (about 100). We might then consider these as a class of agents with continuity in terms of their effect on a sensitive cell, although the same could probably not be said for their evolution. It would seem from their differing modes of action (Table 5) that they have evolved from a wide variety of sources.

Table 5 Modes of action of some colicins.

Col factor	*Colicin*	*Site of action/mechanism*
ColK	K	affects energy dependent reactions in membrane
ColE1	E1	affects energy dependent reactions in membrane
ColE2	E2	endonuclease, degrades DNA
ColE3	E3	prevents protein synthesis by inactivating ribosomes

3.4 Ecology of bacteriocins

The term bacteriocin could encompass a biologically active protein with a molecular weight between 10^2–10^7, although there is no evidence as yet that plasmid borne genes are involved in the synthesis of small peptide antibiotics like cycloserine (MW approx. 100).

What significance do these agents have to their hosts? Do, for example, *E. coli* cells in the intestinal tract have increased survival potential if they are bacteriocinogenic? We are far from being able to answer such questions largely because most data have been obtained through laboratory experimentation

Table 6 A comparison between conditions used in laboratory based studies and those found in the colon.

Laboratory	*Colon*
Generally aerobic	Anaerobic
Nutrient medium	Medium status variable
37°C	40°C
Doubling time estimated at about 30 min	Doubling time estimated at 10 h
	Presence of proteolytic enzymes, bile salts etc. which can all affect the stability of colicins and pili
Mono-culture, only sibling competition. Usually *E. coli* K12 which is acclimatized to laboratory conditions	Competing community

under conditions bearing little relationship to those prevalent in the intestine. Table 6 is an oversimplification of the position but illustrates why the current studies give us a poor insight. The situation is obviously complex but experiments with germ free animals inoculated with Col^+ and Col^- have shown that these strains co-exist. These studies indicate the types and range of defence mechanisms available to individual cells in competitive environments but we still have little idea of their ecological potency.

4 Resistance Plasmids

4.1 The problem posed

Chemotherapy has enabled man to reduce both the fatal and crippling effects which can result from bacterial infections. However, in finding and using these drugs to fight disease it has become clear that bacteria have responded and developed resistance mechanisms. These defence mechanisms can render the cell insensitive to chemotherapeutic challenge and bacterial strains demonstrating a variety of resistance mechanisms have been isolated. Ideally then in the treatment of bacterial infections one must have a clear understanding of the capability of the drug 'in vivo' and a sound knowledge of the bacterium causing the problem. It is invariably the absence of the latter that creates dilemmas in treatment.

Prior to 1943 two theories were advanced to explain the origin of drug-resistant bacteria. One stated that this resistance evolved spontaneously, in the absence of the drug, and was known as the 'mutational theory'. The second stated that resistant cells arose as a direct result of an interaction between the bacterium and the drug. This was called the 'interaction theory'. In 1943, Luria and Delbruck carried out an experiment which supported the first. The evidence clearly showed that drug-resistance variants occurred spontaneously and were present in the population of cells prior to any exposure to the drug. This work established in the minds of many bacteriologists the impression that all antibiotic resistances arose spontaneously. At that time it seemed logical to extrapolate these laboratory studies and conclude that natural bacterial populations would behave in the same manner.

In Japan, about 1945, most outbreaks of dysentery were caused by *Shigella* sp. and the drug sulphonamide (Su) proved to be effective in its treatment. By 1952 more than 80% of the *Shigella* isolates from such outbreaks were highly resistant to the drug. The introduction, about 1950, of the new antibiotics streptomycin (Sm), chloramphenicol (Cm) and tetracycline (Tc) solved this immediate problem. However, after the extensive use of these a few resistant isolates were obtained from cases of dysentery. Initially such isolates showed resistance to only one of the four drugs. In 1956 however, the pattern was found to have changed and one strain was isolated which possessed resistance to all four drugs (Su, Sm, Tc and Cm). By 1957 such multiple-resistant organisms were being isolated regularly and by 1964 about 50% of the isolates were simultaneously resistant to all four drugs. To explain the development of this resistance pattern solely in terms of chromosomal mutation and selection is not realistic because it is a highly improbable event. We can say this because of several observations which suggest another explanation. When laboratory-derived drug resistant mutants are examined they usually show only a slight increase in drug resistance. To obtain the high levels of resistance observed in natural isolates requires the successive accumulation of chromosomal mutations. Such laboratory derived mutants having high levels of resistance are noticeably affected. These mutations have a deleterious effect on the growth rate of the cell; wild type isolates are not similarly affected. Also, to achieve such mutants naturally one would expect them to be exposed to steadily increasing levels of the drug, a factor which is unlikely to occur to a large extent in natural systems. Therefore, since several mutations are required to reach these levels of resistance, and the mutation rate per bacterium is of the order of 10^{-7} to 10^{-10} per bacterium per generation, it seems unlikely that such a progressive type of evolution will occur. While it cannot be ruled out, it does suggest that another and more likely explanation exists.

4.2 Transmissible drug resistance

In Japan epidemiological observations on the changing pattern of antibiotic resistances in strains of *Shigella* revealed some unusual features. It was noted that both sensitive and multiple-resistant strains were isolated from the same patient. These strains were of the same serotype and often the drug-sensitive strains became multiple-resistant as a result of treatment by one drug. This is

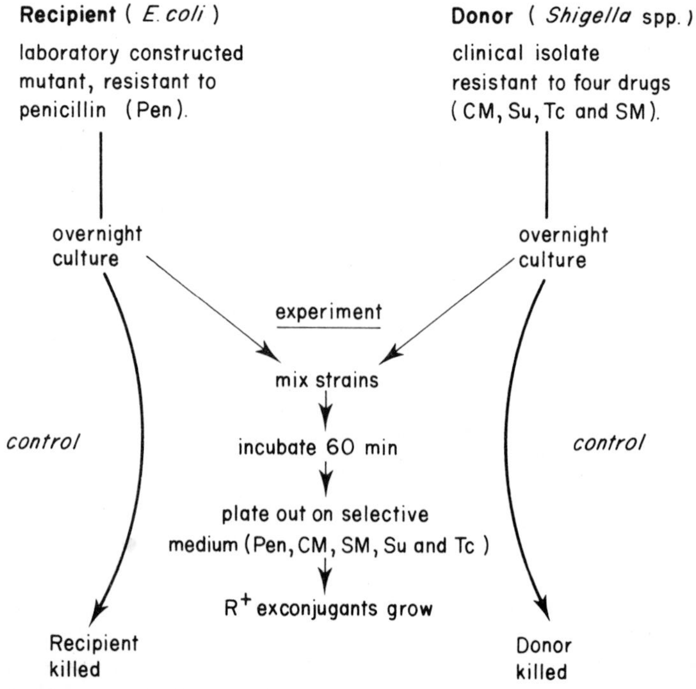

Fig. 4–1 The protocol of an experiment to demonstrate the transfer of multiple drug resistance.

obviously not what would be expected from the mutational theory (see section 4.1). It was further noted that this multiple-drug-resistance pattern was not restricted to *Shigella* spp. but that other bacteria, notably *E. coli* also carried these resistances.

In 1959 two independent research teams, of Akiba and Ochai, presented results which showed these drug resistances could be transferred between bacteria (Fig. 4–1). A laboratory derived mutant resistant to penicillin (Pen) was used as the recipient. The donor was naturally sensitive to penicillin, and resistant to several other drugs. Mixing the two strains resulted in the transfer of the four drug resistances in a co-ordinated or 'all together' manner. Evidence soon accumulated to show that such transfers could occur in the human gut between strains of *E. coli* and *Shigella* spp. Subsequently it was shown that this process was distinctly different from transduction or transformation. Mitsuhashi and Watenabe and co-workers confirmed these findings and extended the work by demonstrating that cell-to-cell contact was required. The process became termed conjugation. The term R factor was proposed for transferable elements carrying drug resistances. By 1963 independent work in Europe had shown R factors to be widespread there and thus they were recognized as presenting a global health problem. The broader and more serious implications for medicine are obvious, in that not only do R factors render the host cell resistant to drugs but they have the capacity to transfer and distribute this capacity throughout the microbial flora. This phenomenon was soon recognized as bearing close relationship to the sex factor described by Jacob and Woolman in 1961. Theirs had been an academic study on genetic systems in *E. coli.*

Since then work has demonstrated many fascinating insights into R factors. Table 7 shows part of the host range for some R factors and the frequency of the

Table 7 Interstrain transfer frequency of R factors (carrying CB resistance) to *P. aeruginosa.*

Donor strain	*Frequency of transfer of carbenicillin resistance* to* P. aeruginosa		
	Rl8–1	R68	Rl8
E. coli	2×10^{-6}	$2 \times 10^{-6+}$	5×10^{-6}
S. flexineri	nd	$2 \times 10^{-6+}$	5×10^{-4}
S. typhimurium	nd	$2 \times 10^{-6+}$	5×10^{-5}
P. aeruginosa	.5	5×10^{-2}	5×10^{-2}

* frequencies achieved by the four donor species.
+ required overnight matings because of low transfer frequencies.

transfer to other strains in the laboratory. The transfer frequencies clearly and principally depend on characteristics possessed by the recipient cell and not on those of the donor species. Figure 4–2 shows that when R factors are transferred in matings where the selection is made for carbenicillin resistance only, exconjugants arise with an unexpected pattern. These are caused by 'breaking up' of the plasmid or its dissociation into smaller entities. Dissociation of plasmid characters is common and presents a problem in their maintenance and analysis.

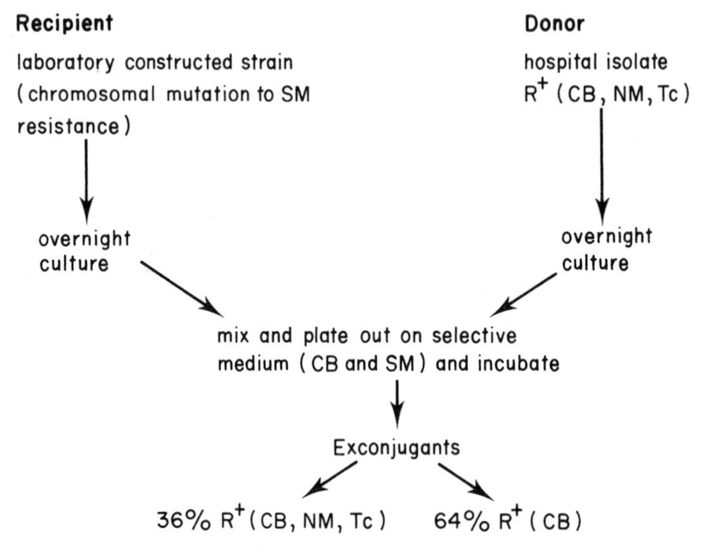

Fig. 4–2 The dissociation pattern of an R factor. R^+ (CB, NM, Tc) = R factor carrying resistances to carbenicillin (CB), Neomycin (NM) and tetracycline (Tc). R^+ (CB) = R factor conferring carbenicillin resistance. The exconjugants are tested on media containing permutations of the four drugs.

4.3 Biochemical mechanisms of drug resistance

R factors carrying genes conferring drug resistances to most clinically used antibiotics have been found. In most cases these have been observed in isolates after the intensive use of the drug. The demonstration of such a widespread distribution of R factors in bacteria has prompted many research workers into attempting to define the biochemical mechanisms of this resistance. The fact that it was a considerable time before it was realised that there was a fundamental difference between laboratory isolated mutants and natural isolates hindered much of the early work in this area. The mode of action of the drug was known before the biochemical bases of the resistance were. Once the mechanisms of resistance were understood there was every chance that a drug could be tailored to be insensitive to plasmid-borne resistance mechanisms.

Spontaneous resistant mutants to streptomycin obtained in the laboratory arise from an alteration in their 30s ribosomal subunit. The mutation modifies the target site and hence prevents the drug from binding. It is usual for natural isolates to show resistance to aminoglycosides by detoxifying the drug, using an enzyme which deactivates the antibiotic. Table 8 gives some other drug

Table 8 The principal plasmid borne mechanisms of drug resistance.

Drug	*Detoxification*	*Alteration in permeability*	*Modification of target site*	*Replacement of target site*
Aminoglycoside[a]	+	+	+	–
β lactams[b]	+	–	–	–
Chloramphenicol	+	+	–	–
Sulphonamides	–	+	–	+
Tetracycline	–	+	–	–
Trimethoprim	–	–	–	+

[a]streptomycin, neomycin, kanomycin, etc; [b]cephalosporin, penicillin, ampicillin, etc.

resistance mechanisms. One affects the uptake rate or permeability of the drug and a second replaces the target site. In the latter the plasmid carries a gene which codes for an enzyme which is insensitive to the drug. It can effectively substitute for the inhibited host enzyme. Table 9 summarizes the biochemical basis for these resistances.

Table 9 Biochemical basis of plasmid borne resistances.

Drug	*Biochemical mechanisms*
β lactams	enzyme cleaves β lactam ring and hence detoxifies the drug
Chloramphenicol	acetyltransferase acetylates and causes loss in activity of the drug
Trimethoprim	dihydrofolate reductase enzyme is insensitive to drug. It takes over metabolic function of inhibited host enzyme
Tetracycline	permeability barrier established preventing drug entry
Lincomycin	alteration of target site (23s RNA molecule) which prevents the drug from binding and inhibiting protein synthesis

4.4 Non-antibiotic antimicrobial agents

Although work on plasmid-borne drug resistance has dominated work on plasmids it is becoming increasingly clear that plasmids carry other types of resistances. These include the wide spectrum of antimicrobial agents shown in Table 10.

Plasmids conferring resistance to many of these agents have been found in both Gram-positive (*Staphylococcus aureus*) and Gram-negative (*P. aeruginosa*) bacteria. In many cases these characters are found associated on the one plasmid. Often these characters are associated with drug resistances. It is clear from the results of large screens that resistances to a whole range of antimicrobial agents occur in the environment. In many cases it may be difficult for us to appreciate the evolutionary steps that have caused these associations, but it is presumably selection pressures (e.g. those that occur in clinical situations) which cause these associations to be observed.

It is just these types of associations that enable the host cell to survive the

Table 10 Plasmid borne resistances carried by bacteria.

Anions	*Cations*
$AsSo_3^{3-}$	Ag^{+}
$AsSO_4^{3-}$	Bi^{3+}
BO_2^{2-}	Co^{2+}
B_4O^{2-}	Cd^{2+}
CrO_4^{2-}	Hg^{2+}
	Zn^{2+}
Organomercurials	*Non-metallic agents*
Phenyl mercury	Hexachlorophene
Methyl mercury	
Ethyl mercury	

challenge of toxic concentrations. Take for example the work carried out on mercury resistance in bacteria. This appears to be located, without exception, on plasmids. In areas of high industrial pollution, high concentrations of organic mercury compounds occur. These have a low toxicity until converted by the bacteria in the sediment into the highly toxic methyl mercury and related compounds. These are potentially lethal to most of the aquatic life forms, including the bacteria carrying out this transformation. This deadly situation can be overcome by plasmid coded enzymes of some bacteria which can detoxify these compounds. It is achieved through two enzymes (see Fig. 4–3).

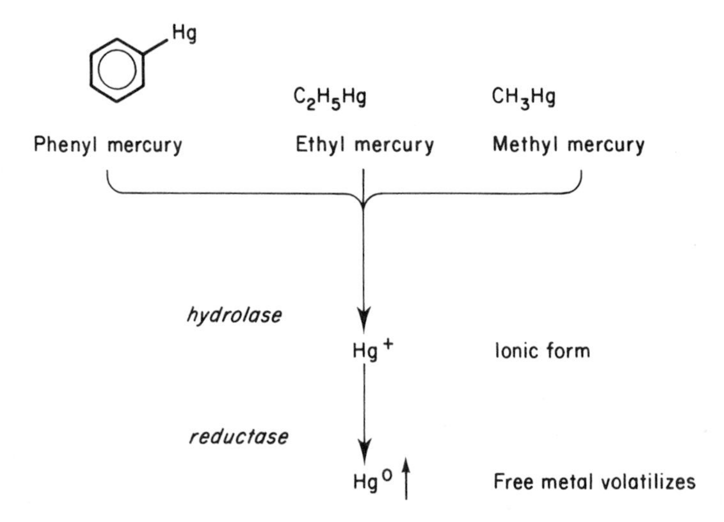

Fig. 4–3 One biochemical mechanism for the detoxification of organomercurials.

4.5 Radiation resistance

In the late 1960s a colicin plasmid, ColIb was shown to partially protect *Salmonella typhimurium* from the mutagenic effects of U.V. light. Since then a number of plasmids with such an ability have been found.

The basis of this protective effect lies not in decreasing the damage to the DNA incurred by exposure to the radiation, as might be expected. It lies, in fact, in an ability to improve the chances of correct repair of the damaged DNA. The plasmid bears an enzyme which can carry out this process. In some plasmids this repair system is also effective against damage caused by chemical mutagenesis. It is probably better to consider plasmids with this capacity as conferring on the cell an increased tolerance to mutation because of increased efficiency of correct repair rather than resistance *per se*.

5 Catabolic Plasmids

In the biosphere we can see highly complex organic compounds are consumed by the higher organisms, readily available nutrients are removed and the remaining material is excreted. Bacteria and fungi have the metabolic capacity to utilize this type of waste as a growth source and enable it to be recycled. This process of utilization is termed *catabolism* or *degradation*. The turnover capacity of the microflora is enormous, as may be appreciated by the rapid disappearance of organic matter. This activity can be demonstrated simply in the laboratory by the addition of soil to a medium containing one of the compounds listed in Table 11. Both communities and individual bacteria responsible for the degradation can be isolated from the growth medium. Indeed such is the capacity of the soil microflora to cope with challenges that, with suitable enrichment conditions, many man-made compounds or *xenobiotics* can become growth sources. When these catabolic capacities are located on a plasmid the latter is termed a *catabolic* or *degradative plasmid*.

5.1 Simple degradative plasmids

The ability of the F plasmid to integrate into the chromosome of *E. coli* has been mentioned (section 1.5.2). Faulty excision of F inserted at a site adjacent to the lactose genes (LAC^+) on the chromosome can result in F′ LAC^+ formation. The genes for lactose metabolism are now carried on the F plasmid. The transfer of such a F′ LAC^+ plasmid to a LAC^- recipient (a cell which cannot grow on lactose) allows the cell to grow on lactose medium. The relative ease with which these plasmid types can be isolated suggest that plasmid/chromosome interactions are not infrequent events. The F′ LAC^+ plasmid is considered a simple catabolic plasmid because it carries about four genes, those for entry and breakdown of lactose in the bacterial cell. Lactose plasmids occur naturally in several genera e.g. Lactobacilli and Streptococci.

5.2 Catabolic plasmids

The occurrence of many complex and highly polymerized compounds in the environment has been mentioned. The degradation of such compounds has been the subject of much interest in recent years. Plasmids have been shown to code for enzymes responsible for the degradative pathways of terpenes, alkaloids, aliphatics and aromatics.

This wide range of catabolic activities coded on plasmids reflects the diverse nature of carbon sources available in the biosphere. On examination these plasmids have all been shown to contain gene clusters which enable a

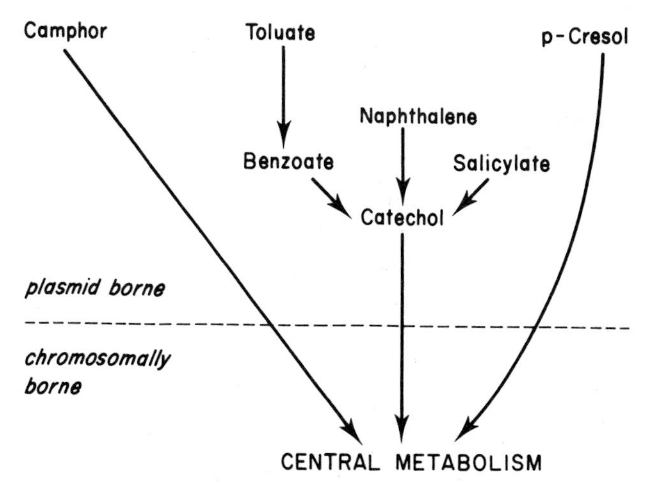

Fig. 5–1 Some of the biochemical pathways present on plasmids in *P. putida*.

coordinated dissimilation of the substrate, feeding the end product neatly into the central metabolism of the cell (Fig. 5–1). This all suggests a high degree of genetic and biochemical precision and co-operation between the plasmid and the host cell. Table 11 gives a list of compounds whose degradation is known to be plasmid borne, although in some bacteria these pathways may be chromosomally coded. These plasmids are often large, 50–90 Md. It is proposed that the

Table 11 Plasmid encoded pathways in *P. putida*.

Plasmid	*Size (Md)*	*Pathway*
OCT	27	octane
SAL	51	salicylate
CAM	92	camphor
NAH	63	naphthalene
TOL	63	toluate
XYL	11	xylene
pND50	nd	p-cresol
MDL	nd	mandelate
QUI	nd	quinate

nd = not determined.

clustering of catabolic genes, to form these catabolic plasmids, occurs randomly by the processes of recombination and transposition (see Chapter 7). The chance occurrence of a certain gene aggregate in a host cell could confer a growth advantage which results in enrichment through selection pressure. Once clustered, such genes can act as an efficient working unit. In matings the recipient acquires a large increase in genetic potential. Often these plasmids possess sex factor activity facilitating their own transfer. The genetic flexibility conferred on the host population from the dissemination of a wide variety of operative gene clusters is obvious. It permits an individual cell the advantage of a small genome and yet allows the population to have a large gene pool, through the maintenance of a large variety of plasmids. This type of operonic clustering is being found in many bacterial species.

5.3 Catabolic plasmids and environmental pollution

A vast array of man-made chemicals are now used in industry and agriculture. While many of these chemicals present no problems, some are recalcitrant and persist for considerable periods of time causing various degrees of pollution. Research into the biodegradability of such chemicals has resulted in some detergents being chemically modified in order that they degrade more readily.

However, some compounds cannot be changed and, after use, persist in the environment. Enrichment studies have shown that for many of these compounds it is possible to isolate bacteria capable of degrading them. Bacteria isolated through their ability to utilize some compounds, e.g. 2.4-D (dichlorophenoxyacetate), carbaryl and polychlorinated biphenyls, have been shown to possess plasmids concerned with the degradation. This finding shows clearly that bacteria possess the capacity to evolve novel gene arrangements to meet the changing environmental challenges.

Oil spills and concomitant pollution are a common feature of industrial areas and a proposal for using bacteria to clean up such 'spills' has been made. A patent for a laboratory-constructed pseudomonad containing the plasmids NAH, CAM/OCT and TOL has been granted. It has been suggested that this bacterium would be able to utilize, and grow on, the various oil components thus removing the pollution. While the idea is commendable, the 'heavy oil' fraction is always the one which remains and this is the least likely one to be utilized by the bacteria. Even so, it is probable that, in the near future, constructed strains will be utilized to 'clean up' areas contaminated with chemicals.

6 Plasmids involved in Pathogenicity

Pathogenicity describes the ability of an organism to produce disease. *Virulence* is a term that refers to those additional properties that enhance pathogenicity and is one measure of the potential seriousness of the infective agent. We refer to disease producing organisms as pathogens, but it is becoming increasingly clear that many micro-organisms which show pathogenic properties only do so because they carry a particular plasmid.

6.1 The human host

Table 12 shows some plasmids which confer pathogenicity on a host bacterium and cause disease in humans.

Table 12 Some plasmids which endow their host cell with pathogenicity.

Organism	*Disease*	*Causitive agent*
Streptococci Group A and B	Scarlet fever	phage lysogen
Clostridium botulinum	Botulism	phage lysogen
C. diphtheriae	Diphtheria	phage lysogen
E. coli	Bacteremia/ meningitis	a variety of plasmids
Staphylococcus spp.	Diarrhoea	a variety of plasmids

6.1.1 Diphtheria

Diphtheria is a serious disease caused by *Corynebacterium diphtheriae*. It infects the upper respiratory tract and produces a toxin which, when adsorbed, damages the heart, kidney and nervous system. This powerful toxin is produced by lysogenic strains only. Cured strains, those which have lost both the virus and the ability to produce the toxin, regain this ability on becoming reinfected by the virus.

6.1.2 Diarrhoea

This is caused by several enteric bacteria, viruses and protozoa. Recent research has indicated that more than 50% of cases are caused by *E. coli*. This is surprising because this organism forms such a significant proportion of the intestinal population. It appears, however, to be only those *E. coli* strains possessing a particular type of plasmid which can cause the disease. These 'converted' strains invade and grow in the intestinal mucosa of the small bowel. The bacterial cell produces a plasmid coded toxin which causes inflammation and stimulates the adjacent tissue to secrete fluid, inducing the classic symptoms of diarrhoea. Early work, carried out in piglets, showed that pathogenic strains possessed a plasmid which produced a protein, antigen K88, identified as a pilus. The pili enable the cells to adhere to the wall of the small intestine at sites optimal for the adsorption of the toxin.

6.2 The plant host

The Gram-negative organism *Agrobacterium tumifaciens* has interested plant pathologists owing to its *oncogenicity*, or tumour inducing capacity, both in

Fig. 6–1 A crown gall tumour. (Courtesy Dr J. Schröder, Max-Planck Institute, Cologne.)

dicotyledons and gymnosperms. Such tumours, also known as crown galls (Fig. 6–1), were recognized in 1907 as being caused by an organism then called *Bacterium tumifaciens*.

The process of tumour formation, termed *transformation*, describes the conversion of a plant cell into a cancerous cell capable of autonomous growth. It has a distinctly different meaning as used here from that understood by microbial geneticists. The process of transformation is shown in Fig. 6–2 and

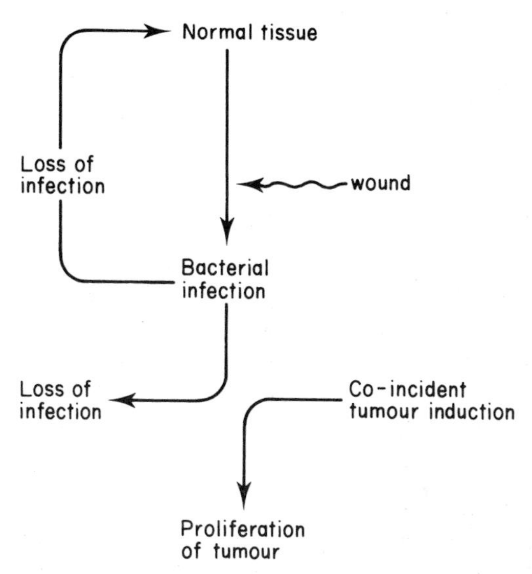

Fig. 6–2 The phases of tumour formation.

starts with the infection of a wound which permits the bacteria to invade the intracellular spaces. These bacteria do not appear to invade the cells but from this position proceed to convert (a plant cell can be transformed from a normal cell by an outside agent) the adjacent cells over the next 3–4 days. Once the cell has been converted it grows independently of the surrounding host cell tissue forming a tumour. The formation of this tumour is dependent on a temperature below 30°C, and once formed it is self-proliferating, graftable and it can also be maintained bacteria-free in the laboratory.

Oncogenic bacteria have been found to contain a plasmid, a Ti or tumour inducing plasmid, about 120 Md in size. The loss of this plasmid removes the ability of the host cells to form tumours. The ability is returned by re-introducing the plasmid. Subsequent work has clearly shown the essential role the Ti plasmid plays.

Investigations have shown there to be two types of Ti plasmid, differentiated by their ability to catabolize the unusual amino acids nopaline and octopine. These plasmids often possess other genetic abilities e.g. genes for conjugation (Tra) and oncogenicity (Onc). Taken at face value the association of these characters may seem old. However, once it was realised that the tumours themselves produced either nopaline or octopine the situation became clearer. It seems that the resident bacteria, carrying the Ti plasmid, induce the plant cell to synthesize one of these two amino acids which are then utilized by the bacteria for growth. The situation is more complex because it appears that the bacteria also induce changes in the processes which regulate cell division, hence the tumour.

Another type of tumour is formed on the roots of legumes by an infection of *Rhizobium* sp. This tumour, known as a *nodule*, fails to develop in the same spectacular manner as a crown gall, but covers the roots with tiny nodules. Again the bacteria invade the tissues and induce a local proliferation of tissue which surrounds the infected area. The bacteria undergo a metamorphosis and become bacteroids. Such cells have lost their growth autonomy, becoming dependent on the host for growth; they swell up to forty times their original size and form characteristic 'X' or 'Y' shaped cells. These bacteriods fix nitrogen gas into nitrogen compounds which are used by the plant. Both the ability to nodulate and the choice of host plant are specified by the plasmid. This can be shown by curing one strain and reinfecting it with a plasmid from another unique strain. The reinfected strain adopts the new infective and nodulating characteristics of the unique strain. It has now been demonstrated that the genes for nitrogen fixation are also *borne* on such plasmids.

This is an exciting area of research for several reasons. Firstly it demonstrates that a normal plant cell can be converted by an outside agent into a cancerous cell. At present the evidence suggests that this conversion results from the incorporation of DNA transferred from the bacterium. The genes are stably maintained and expressed in the host plant cell. This conversion does not appear to result from the synthesis of some type of 'messenger' signal which switches on certain genes already present in the cell. If bacterial genes are incorporated then it should soon be possible, with the current capabilities in molecular genetics, to define the genes involved in the process, how they leave the bacterium and enter the plant cell. It may then be possible to analyse the processes involved in the conversion of a normal cell to one in a cancerous condition. There are other potential benefits. From such a study of cellular regulation we will learn more about plant growth and consequently may be able to improve growth efficiency and hence agricultural productivity. Once the mechanisms of transformation of the plant cell by bacteria involved in the passage of DNA from the bacterium to the plant cell are understood, it may prove possible to modify them genetically and develop new areas in plant genetics. For example, it has been proposed to transfer the nitrogen-fixing genes to non-leguminous crops. This would obviate the need to add nitrogen-based fertilizers. At present we do not know what effects, if any, this suggestion would have on either the crop yield or the ecosystem.

6.3 The bacterial host

Bacteria are infected by bacteriophage and these infections, sooner or later, end in the death of the host as a result of multiplication and the release of the progeny. There is a great diversity and abundance of bacteriophages and they have an obvious predator-prey relationship with bacteria. Despite this it seems that in some instances there is a net benefit to a host cell from non-lethal infection.

6.3.1 Beneficial effects of virus infection

One can envisage situations in which the presence of a virus could be considered as conferring a net benefit upon a host cell. However, to appreciate this argument one has to consider the advantages to a population of cells and not to individual members of the population. In one situation the virus would be capable of transducing genes between members of the population (see section 1.5.1). If these genes are of value to the receiving cell and enable it to survive some challenge then a net benefit will have been derived. It may be possible for the genes to be stabilized by integration. Alternatively the host cell may be lysogenized (see section 6.1). This results in an alteration to the properties of the outermost layers of the cell containing the virus receptor sites. This effectively immunizes the cell from infection by other viruses or sex factors utilizing the same receptor sites. It is also possible for such alterations to have other effects e.g. changing the permeability, hence sensitivity, of the cell to drugs.

6.3.2 Predator-prey equilibria

Without the occurrence of mutations to resistance or the phenomenon of lysogeny, viruses would be expected to wipe out a population of sensitive cells.

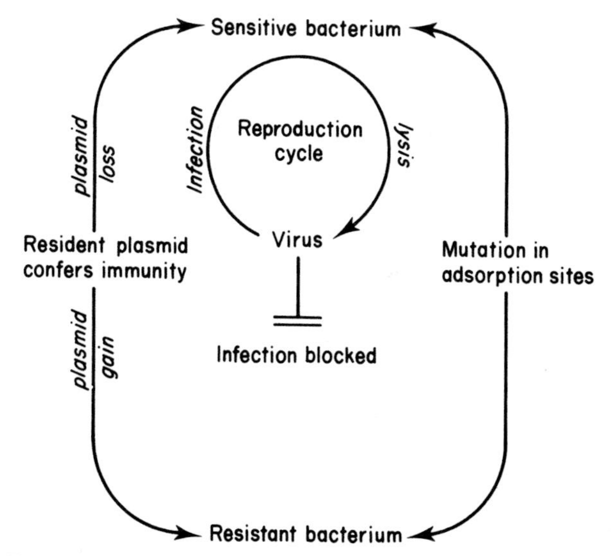

Fig. 6–3 The basis of the dynamic equilibrium between bacteriophage and their host bacterium.

The dynamic equilibrium between lysogeny and mutation to resistance can be demonstrated in the laboratory. Since both types of resistance are reversible the evolution of virus-sensitive populations can arise. Figure 6–3 shows some of the possible relationships. The situation is effectively a dynamic one because it is possible for a virus to change its adsorption sites on the host by mutation or through recombination with a different virus. This results in the infection of previously immune cells. The presence of a resident plasmid, e.g. a prophage or R factor, can also confer immunity to infection.

In conclusion the interaction between bacteriophage and their hosts is characterized by several features which are analogous to those demonstrated by classical plasmids.

7 Evolutionary Mechanisms of Plasmids

It is reasonable to ask how these extrachromosomal elements have evolved and why they should be maintained in a bacterial host. Plasmids, like other genetic elements, evolve so as to maximize their numbers. Presumably, therefore, innovations which permit them to do so and spread more efficiently will be selected and become established.

7.1 Plasmid stability

The stability of a plasmid is measured by the persistence in time of some or all the constituent characters held on the plasmid. Plasmids with low stability will be totally or partially lost from individual cells in the population as affected by curing or dissociation. Both these events are influenced by many parameters (Table 13). There is much interest in the factors affecting this stability and current work suggests (Table 14), as may be expected, that the situation is highly

Table 13 Factors affecting plasmid stability.

The host	Restriction systems
Growth rate	Growth nutrients
Temperature	Other plasmids
Radiation	Chemicals

Table 14 The stability of the R factor R18–1 in different hosts after 15 generations in non-selective medium.

	% R^+ strains	
Host	*0 generation*	*+15 generations*
P. aeruginosa	100	100
E. coli	100	99
S. flexineri	60	1

complex. It is based upon interrelated parameters, e.g. selection pressure, temperature and nutritional status. It also has a statistical parameter because of the random nature of the events which culminate in dissociation. Often plasmids are quite stable, i.e. no observable curing rate, but for others dissociation/loss occurs at 10^{-4} generation^{-1} and higher. Such results reflect the stress which plasmid/cell associations are under.

7.2 Insertion sequences and transposons

The bacterial chromosome consists of a sequence of genes linked in the form of a circle. It is a common concept that this sequence is rigidly defined. Genes can leave and re-enter the sequence at new sites and a mechanism for this movement, termed *transposition*, has been discovered. This genetic ability to reorganize the genome gives us an idea of how the chromosome may have evolved. Genes which have a common function could become aggregated into clusters. These could then become co-ordinately regulated into functional and efficient units. Gene clusters have been observed in many bacteria.

What is the mechanism of transposition? Bacteriophage λ (MW 48 000) has one usual site of entry in the chromosome, but can very occasionally enter at different sites. It is, therefore, considered to be site specific. Bacteriophage Mu–1 (MW 38 000) is different, it randomly inserts and excises itself from the chromosome. As Mu–1 'moves about' it is in effect *translocating* virus genes about. There are other elements which exist in the cell which can move about between plasmids and chromosome. These are termed IS or *insertion sequences* and are discrete genetic and physical structures. Their movement is determined by a recombination system which is specific to these elements and which recognizes the ends of the elements for the insertion or excision steps. The enzyme system used is different from the one used by the cell for normal recombination (*rec*A). In the process of integration or excision of λ there is no DNA synthesis; it is a normal type of recombination.

It appears that during the process of transposition of Mu–1, transposons and IS elements, the genetic material concerned duplicates itself in such a way that all, or part, remains at the original site; *rec* A independent translocations can best be thought of in these two categories.

7.2.1 Insertion sequences

IS elements appear to be a normal constituent of the chromosome of *E. coli* and other bacteria. *E. coli* K12 has multiple copies of IS1 (8), IS2(5) and IS3(3) elements in its chromosome, equivalent to about 1% of the genome. Subsequent studies have demonstrated that these and other IS elements occur in the genomes of other bacterial species. They are from about 700 to greater than 2000 base pairs in size.

It has been estimated that 10–15% of all spontaneous mutations (e.g. those to auxotrophy) are caused by these elements inserting into a gene, making it inactive.

7.2.2 Transposons

Drug resistance genes can be shown to 'move about' the genome of *E. coli.* These entities are larger than IS sequences, e.g. Tn1 codes for ampicillin (Amp) resistance (MW = 4500) and Tn4 codes for Amp, Su and SM (MW = 20 000). Transposons have been shown by heteroduplex analysis to carry unusual

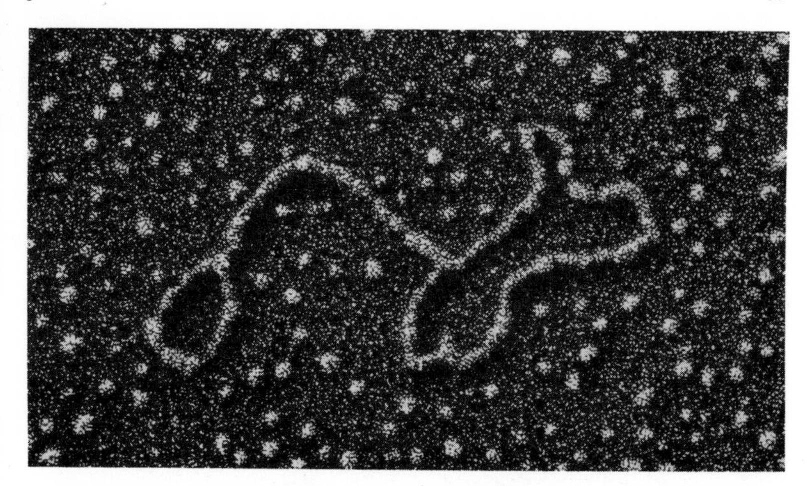

Fig. 7–1 An electron micrograph of single-stranded plasmid DNA showing a snap back structure typical of many transposons. The large loop is the replicon into which the transposon has inserted. The stem represents double-stranded DNA formed by the annealing of an inverted repeat sequence which flanks the genes in the small loop. It is this linear stretch plus the small loop which represents the transposon. (Courtesy of Dr S. Cohen, Stanford University Medical Centre, Stanford, California.)

structural features; the two ends have stretches of homologous base pairs forming insertion sequences (Fig. 7–1). These two IS elements co-operate to mediate the translocation of the intervening genetic sequence. A route exists by which highly complex translocatable elements, e.g. R factors, can be assembled (Fig. 7–2). In fact even the virus Mu–1 can transpose DNA if it is held between two Mu–1 genomes. The removal of genes from the chromosome by translocation has a second effect; it causes deletions of the intervening and adjacent genes thus generating new gene sequences.

7.2.3 Bacteriophage

The two bacteriophage mentioned, λ and Mu, insert their genomes into the chromosome independently of the *rec*A gene system. These and other viruses clearly fall into the realm of translocatable elements. It has been estimated that 2% of spontaneously derived auxotrophs in Mu lysogens arise because the phage Mu inserts inside a biosynthetic gene. Table 15 shows that these transposable elements differ in their ability to 'move about'. This frequency may also be influenced by their host.

7.3 Plasmid size

Bacteria exist naturally in communities, although there are exceptions in extreme environments, e.g. the acid-tolerant thiobacilli which are considered as

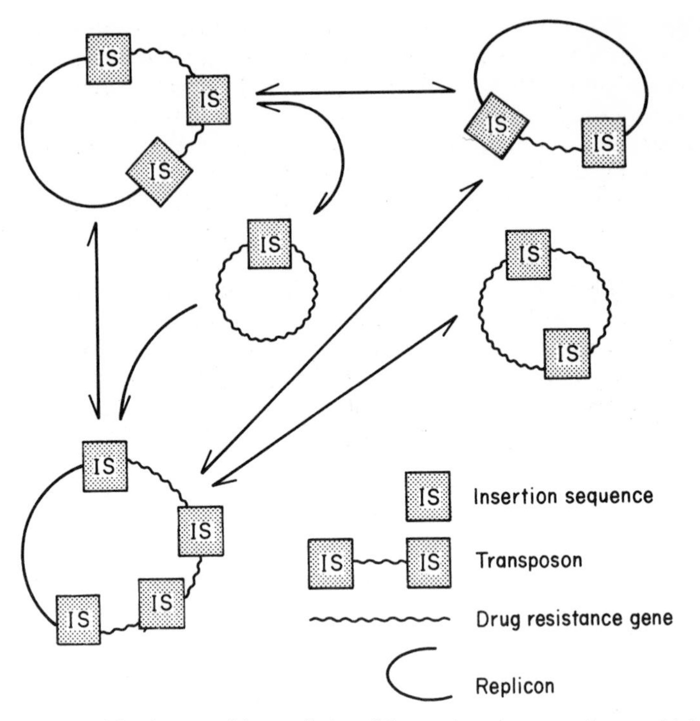

Fig. 7–2 A model for the reversible association of drug resistance genes to form multiple-drug-resistant plasmids (R factors).

Table 15 Estimates of translocation and deletion frequencies per cell duplication.

Genetic element	*Translocation frequency*	*Deletion*
λ	inefficient (⊳ 10^{-6})	10^{-7}
Mu	very efficient (10^{-4})	$\simeq 10^{-6}$ excision $\simeq 10^{-1}$ per integration
IS	equivalent to spontaneous mutation rate (10^{-7})	10^{-4} to 10^{-5}
Tn	variable (10^{-1} to 10^{-7})	probably similar to Mu but frequencies will vary for each transposon

a single population. The community is therefore composed of individuals that interact positively or negatively depending on the prevailing conditions. The stability of the community and its survival are, therefore, linked to factors like the longevity of its members, its resistance to physical stress and its genetic potential. The members are continually tested against this fluctuating environment and, to survive, must have the capacity to respond successfully. One strategy for success would be to accumulate genes of survival value. However, most bacteria have small genomes. How can this be accounted for? As the amount of DNA increases the amount of energy expended to replicate and maintain it increases, thus the efficiency of the cell decreases although its ability to survive rare challenges may increase. Bacteria have utilized plasmids to increase their genetic potential without burdening themselves with a vast amount of extra DNA. The presence of plasmids adds, to the genome, from 0.5% to as much as 30% in extreme cases. *Pseudomonas* spp. and *E. coli* have adopted a strategy of placing genes of transient, but vital, value upon plasmids. The amount of extra DNA any one cell carries is therefore small, but since the population carries a diverse collection of plasmid types the potential of the population is greatly increased. We know that plasmids are a common feature of bacteria and could therefore be capable of disseminating genes through a community. Recent research indicates that plasmids allow characters to be combined and tested in a range of organisms. This implies that plasmid evolution is a continual and dynamic process.

7.4 Origin of phenotypes

Where do the genes come from? The problems encountered in medicine with drug resistant bacteria have been well publicized, and Table 1 indicated the spectrum of properties carried. When examined, plasmids seem to carry more than one property and it appears that there is selective advantage accruing to the

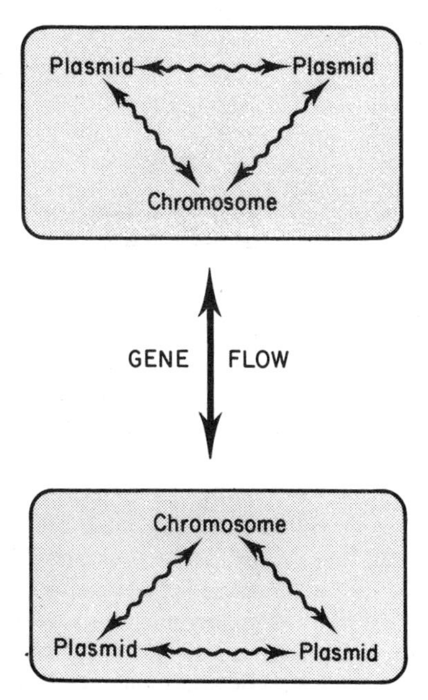

Fig. 7–3 The processes involved in gene flow in bacterial communities.

host by the acquisition of such combinations. Take for example the aggregation of drug resistance genes; there is no evidence for these having evolved on the chromosome of enterobacteria, but there is evidence for them having evolved in unrelated organisms. IS elements are important in gene movement between plasmids and chromosomes. If we add this factor to the mechanisms of gene flow we can envisage how genes can be transferred through a community (Fig. 7–3).

Through this system we can see how characters of intense survival value could transpose from a plasmid into the chromosome to be stabilized. This also suggests that bacteria form a genetic continuum and could explain how genes, e.g. those for resistance to streptomycin, spectinomycin and gentamycin, have apparently 'hitched a lift' through into the coliforms. It is postulated that these resistances were derived from streptomycetes, organisms which naturally produce these antibiotics.

7.5 Host range

Fertility factors may have limited host ranges (e.g. F) or they may be promiscuous (e.g. RP1). R factors with no sex factor activities can also have narrow and broad host ranges. There is therefore a host range spectrum and plasmids like F and RP1 occupy extreme ends (Table 16). Although little work has been done on the host range of bacteriophage a similar pattern is apparent.

Table 16 The host range of some plasmids.

Plasmids	*Host range*
RP1 (R factor)	A wide range of Gram-negative bacteria
F (sex factor)	A few strains of *E. coli*
P1 (virus)	*Escherichia, Salmonella, Shigella* and *Klebsiella* spp.
MS2 (virus)	Will only infect F^+ cells

7.6 Relationships between plasmids

An intracellular distinction between plasmids is difficult to make as they are all *CCC DNA*. It is only when the genes are expressed that they can be easily separated into groups. This similarity in their general structure implies a continuity in their evolution.

For example, the sex factor F promotes its own transfer via a pilus to a recipient cell. There are viruses that will only infect cells carrying the F factor and two of these are MS2 (male specific RNA virus) and f1 (a filamentous virus). These attach along the length and at the tip of the pilus respectively. These two viruses are reliant on the host cell maintaining this particular plasmid for their survival since other types of pili are of no value. The F^+ cell has responded to this by repressing the continual synthesis of the pili and now can only be 'preyed upon' while sexually competent. Also many filamentous viruses, e.g. f1, are not obliged to lyse their host cells, but can extrude themselves in a manner analogous to the production of pili.

If one considers the overlap in other characteristics like compatibility, fertility, inhibition and superinfection immunity then these elements do appear to form a continuum.

7.7 The origin of plasmids

A plasmid is an autonomous genetic unit. It has two principal components; one is concerned with its replication and the second endows the unit with a set of characteristics (Fig. 7–4). The ways and mechanisms by which such elements arise, and are still evolving, are currently the subject of much speculation.

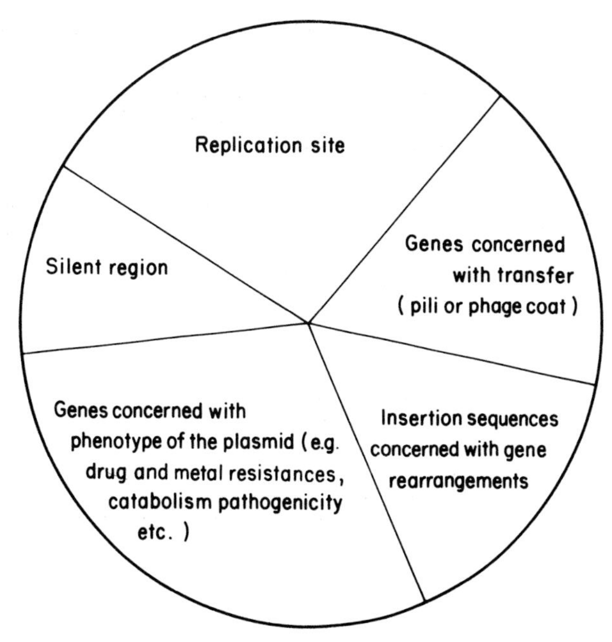

Fig. 7–4 A composite of characters which could be present on a plasmid.

In the evolution of a plasmid one function is essential, the ability to replicate autonomously. At some point in time, possibly before the mechanisms of chromosome segregation were as well organized as now, a mini-chromosome arose by mutation, transposition or an illegitimate recombination or replication error. In this way a replicon was formed. Providing it could be stabilized in the host cell and avoid competition with the resident chromosome it is conceivable that it would be maintained in the cell. The subsequent events of transposition and recombination and mutation could displace characters from the main chromosome to this secondary element. It would then be analogous to plasmids as we recognize them.

For this unit to become transmissible it requires other genes for the synthesis of a viral coat or conjugation tube. We have seen that there is a distinct relationship between these, but it is difficult to see at present how these could have evolved, other than by recruitment from other unrelated functions. Perhaps they are derived from fimbriae-like proteins which have affinity for the membrane of the cell. A mutation may have permitted them to polymerize in an alternative spacial manner. Once formed these could then permit the flow of genetic information and then be selected for because of this characteristic. In support of this mechanism it has been shown that a wide range of enterobacteria, including *Escherichia coli, Shigella* sp., *Salmonella* sp., *Serratia* sp. and *Klebsiella* sp. have a common lipoprotein in their outer membrane. This is a structural feature, or adsorption site, which would allow a broad host range plasmid to identify host cells.

Investigations into plasmids have shown that one drug resistance gene can be present upon unrelated plasmids and conversely one replicon can carry a wide range of unrelated genes. Also many of the smaller plasmids are present in the host cell in multiple copies. It may be that this amplification in copy number increases the chances of producing useful gene combinations. These findings suggest a flux whereby genes are continually picked up and unloaded. Furthermore, it appears that there is a pool of basic replicons in enterobacteria which are involved in this process. It may be that plasmids are most stable in this form.

8 Plasmids and Applied Biology

The previous chapters have illustrated that plasmids are intimately linked to the survival and evolution of bacteria. The aim of this chapter is to reveal their importance to some areas of applied biology.

8.1 Bacterial classification

As taxonomic methods have improved it has become evident that bacteria do not occupy the neat and discrete groups defined by taxonomists. Recent tests have shown that fresh isolates form part of a continuum, many possessing characteristics placing them as intermediates between classically defined species. Previously these have been described as atypical forms. This phenotypic variability clouds the peripheral areas of species identification. Where identification is dependent upon a few key characters, as occurs in hospital diagnostic work, it is important that these characters should be unique to the organism and not occur on, or be transferred by, plasmids. Table 17 lists some genes which

Table 17 Genes which have been acquired by *E. coli.*

	Donor organism				
Characteristic	Salmonella	Proteus	Klebsiella	Citrobacter	Enterobacter
Adonitol utilization	+	+	+	+	−
Citrate utilization	−	−	+	−	−
Lysine decarboxylase	+	−	+	−	+
Phenylalanine deaminase	−	+	−	−	−
urease	−	+	+	−	−

The tests used are adopted for use as indicators and recorded in the following manner: $+ = \geqslant 80$, $- = \leqslant 20$, and $d = 21 - 79\%$. *E. coli* is negative for these characters.

have been transferred to *E. coli* in the laboratory. The acquisition of these in nature would confuse the taxonomist. As there is no reason to suspect this to be a unique situation other identifications could also be at risk.

8.2 Hospital acquired infections

Infections caused by gram-positive bacteria are now successfully treated by the general practitioner. Partly as a result of this it is infections by Gram-negative bacteria that are now the most serious problem, particularly in hospitals. Factors contributing are the change in type of patient and the use of a wide variety of immunosuppressive and cytotoxic drugs. Patients with reduced resistance are present for long-term treatment and these factors render them susceptible to invasion by opportunistic pathogens. Healthy hospital personnel also have a high probability of acquiring such infections and the principal organisms seem to be the species of *Escherichia, Proteus, Klebsiella, Enterobacter* and *Pseudomonas*. These organisms all form part of a microflora which is continually subjected to challenges from antibiotics. This ensures selection for a microflora carrying resistance genes on plasmids. The net result is a reservoir of opportunistic bacteria carrying a wide range of plasmids concerned with their survival and pathogenicity. These factors increase the patient's chance of developing a secondary infection and complicates subsequent treatment.

8.3 Plasmid gene transfers in the medical environment

It has been known from laboratory experiments that *Staphylococcus aureus* bacteriophage can mediate transduction, but it had been thought to be of little significance in natural populations. This was because the 'in vitro' rate of drug resistance transfer was 10^{-6}–10^{-8}. In 1971 transfer of drug resistances was observed to occur on the skin of human volunteers. It now seems that this gene transfer requires direct cell-to-cell contact and what is being seen can best be described as phage mediated conjugation. The virus remains attached to the outermost layers of the bacterial donor cell and until contact is made with a recipient. It is different from conjugation because the DNA is outside the cell in the bacteriophage when contact is made. The clinical situation is further complicated by the fact that these bacteria can also take up plasmids by transformation. If drug resistance markers can be transferred then presumably other characters such as virulence could also.

There is now convincing evidence that drug resistant factors and enterotoxin plasmids etc. are maintained, and capable of transmission, in the gut. Table 6 (page 20) shows conditions prevailing in the gut. These, it has been argued, would be conducive to a low rate of gene transfer for *E. coli*. However, conditions for gene transfer by *Bacteroides* spp. may be ideal. These organisms are obligate anaerobes and can form $> 90\%$ of the gut flora. Drug resistant plasmids are found in these organisms and can be transferred to *E. coli* and other members of the flora. It seems likely that when the total flora is examined the amount of gene flow may be significant.

We know that without selection pressure plasmids tend to dissociate and reform into other combinations. It is not, therefore the drug resistant elements that should be of prime concern, since they are only a transient manifestation of the changes. It is the insertion sequences and the replicon unit which are the prime synthetic agents in the construction of the drug resistance factor. If these two components were examined we might well find a higher degree of gene flow than currently proposed.

8.4 Plasmid involvement in soil and water systems

Many soil bacteria contain plasmids which enable them to degrade a wide range of aromatic compounds. Some plasmids for example code for the breakdown of complex and recalcitrant compounds such as lignin and cellulose. Other more exotic plasmid-borne characters have also been discovered, e.g. antibiotic production in *Streptomyces* spp., the salt tolerance of some marine spirilla and the genes for the synthesis of a gas vacuole and pigments in *Halobacterium halobium* (in this case the buoyancy and pigments interact to optimize the light intensity). This is a good illustration of how genes of unrelated function can become associated, on a plasmid, into a genetic unit of significant survival value.

While plasmids are well documented in Gram-positive species few are capable of mediating conjugation and none has been found with this capability in the genus *Bacillus*. It was suggested that this reflects a fundamental ecological difference between the habitats of coliforms and soil bacteria. For example, in the gut large populations exist and the chances of matings are high. Conversely in the soil, a particulate and colloidal system of variable pH, temperature and

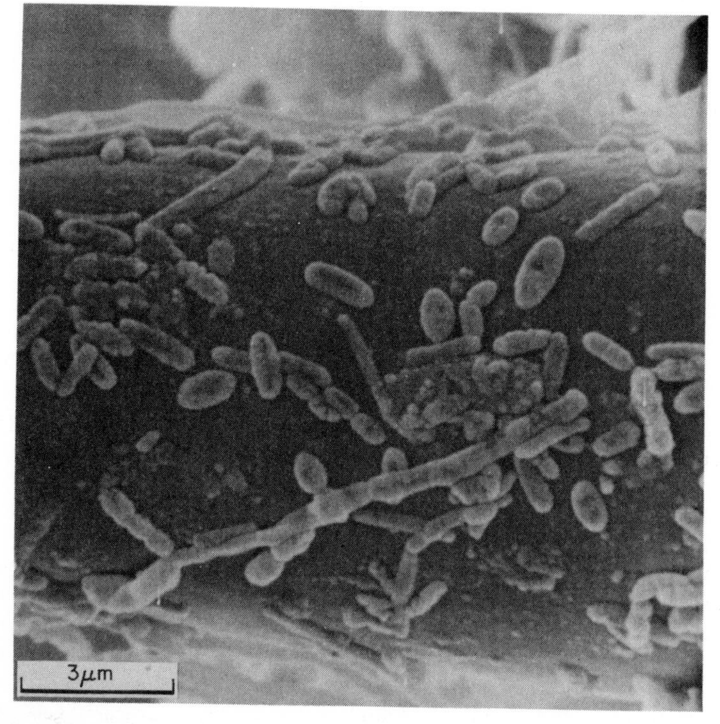

Fig. 8–1 Bacteria on a leaf papilla of water crowfoot. (Courtesy of Dr J. Baker, Freshwater Biological Association, Dorset.)

nutrient supply exists reducing the chances of cell-to-cell contact and subsequent mating. This, however, is too simple an explanation since there are many bacteria which have no association with the alimentary tract and yet can be shown to carry out conjugational gene transfers, e.g. bacteria of skin, plant tissue, sediment and soil crumbs. In fact growth of bacteria in natural systems is often associated with surfaces in the form of microcommunities in a film (Fig. 8–1). In such a film, bacteria are in very close contact and the physical conditions are overtly suitable for gene transfer.

9 Summary

A range of examples has been used to illustrate the role of plasmids in the life cycle of the bacterium. It is beyond the scope of this book to have considered in any detail the use of plasmids in the area of genetic engineering and biotechnology and these aspects should be the subject of a book in their own right (see SMITH, 1981). The same is true for eukaryotic plasmids. Future work will no doubt reveal that these have a significant role in the genetics of higher organisms, including Man.

From the diversity of plasmid types that exist it is now possible to comprehend the extent of the interactions that occur between plasmids and the chromosome of the bacterial cell. Evolution occurs through the selection of genes and gene combinations that confer survival advantage on individual cells within a community. Short generation times, apparently high mutation rates and large populations have been proposed as adequate for bacteria to overcome environmental stress. This concept was derived from laboratory experimentation. A simple comparison of drug-resistant isolates from nature and the laboratory reveals that natural populations do not evolve by mutation alone. This ability to acquire genetic information horizontally from related and unrelated members of the community (Fig. 9–1) means that bacteria have been

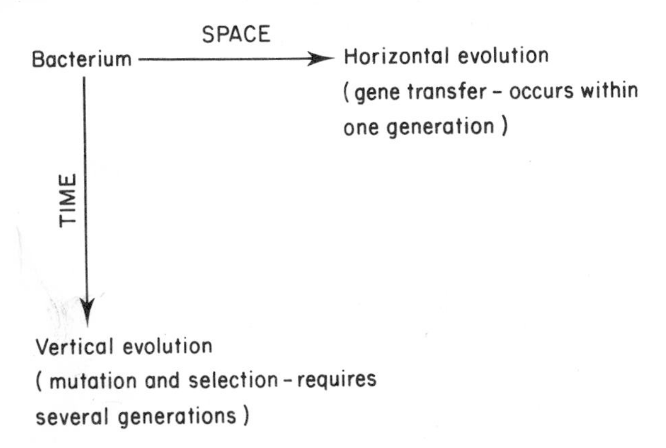

Fig. 9–1 A summary of the two major processes in bacterial evolution.

able to conserve their small genomes and high growth rates. Bacteria therefore become adapted in two ways (Fig. 9–2), they can acquire genes from another source (horizontal evolution) or can change their DNA by mutation to permit growth (vertical evolution).

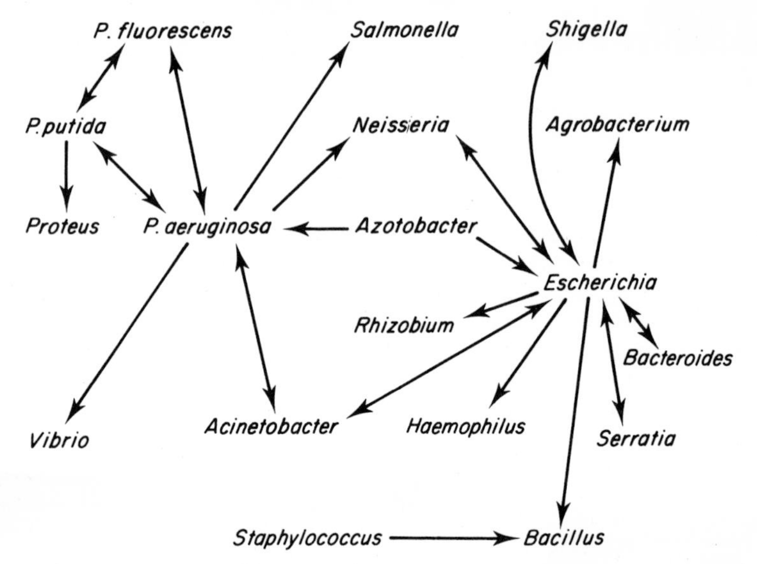

Fig. 9–2 Some examples to illustrate horizontal gene flow between bacteria. The arrows indicate the direction of gene flow recorded.

The thesis of this text is that plasmids together with mutation play a central and fundamental role in the processes of evolution in bacteria. They allow genetic plasticity and confer sexual ability on a group which should not be considered as asexual. It can therefore be seen that plasmids have a central role in the life cycle of bacteria.

Further Reading

BRODA, P. (1979). *Plasmids*. W. H. Freeman Ltd., London.

BUKHARI, A. J., SHAPIRO, J. A. and ADHYA, S. L. (1977). *DNA Insertion Elements, Plasmids and Episomes*. Cold Spring Harbour Laboratories. (A more complicated text for the advanced student.)

COHEN, S. N. and SHAPIRO, C. J. (1980). Transposable genetic elements. *Scientific American*, **242**, 36–45. (A basic introduction to plasmids.)

FALKOW, S. (1975). *Infectious Multiple Drug Resistance*. Pion Ltd., London. (A good introduction to the medical side of plasmids.)

MITSUHASHI, S. (Ed.) (1977). *R Factor. Drug Resistance Plasmid*. University Park Press. (For the advanced student.)

NOVICK, R. P. (1980). Plasmids. *Scientific American*, **243**, 77–90. (A good introduction to plasmids.)

SMITH, J. E. (1981). *Biotechnology*. Studies in Biology no. 136. Edward Arnold.

STUTTARD, C. and ROZEE, K. R. (1980). *Plasmids and Transposons*. Environmental effects and maintenance mechanisms. Academic Press. (Short chapters of recent research and good starting point for advanced students.)

TIMMIS, K. and PUHLER, A. (Eds) (1979). *Plasmids of Medical, Environmental and Commercial Importance*. Elsevier, Amsterdam. (For the more advanced student.)

WATANABE, T. (1963). Infective heredity of multiple drug resistance in bacteria. *Bacteriological Reviews*, **27**, 1, 87–115. (A fascinating review of the early work.)

WILLIAMS, P. A. (1978). The biology of plasmids. In *Companion to Microbiology* (Eds A. T. Bull and P. M. Meadow), Longmans, London, 77–108. (A good introduction to plasmids.)

Index